Ethnobiology

Series Editors
Robert Voeks
John Richard Stepp

More information about this series at http://www.springer.com/series/11551

Aung Si

The Traditional Ecological Knowledge of the Solega

A Linguistic Perspective

Springer

Aung Si
School of Languages and Linguistics
University of Melbourne
Parkville, VIC, Australia

Ethnobiology
ISBN 978-3-319-24679-6 ISBN 978-3-319-24681-9 (eBook)
DOI 10.1007/978-3-319-24681-9

Library of Congress Control Number: 2015952625

Springer Cham Heidelberg New York Dordrecht London

Printed on acid-free paper

Springer International Publishing AG Switzerland is part of Springer Science+Business Media (www.springer.com)

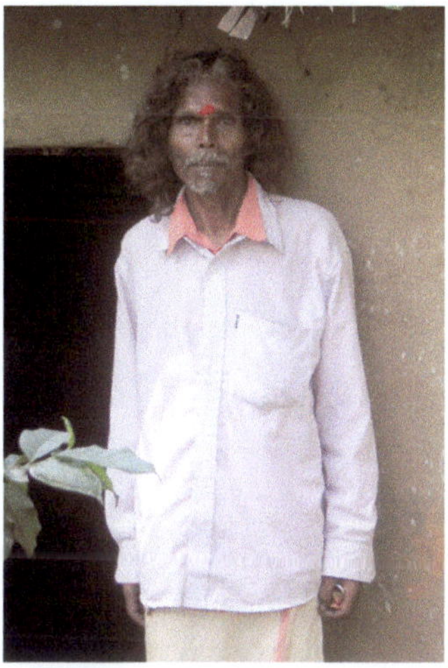

This book is dedicated to the memory of Heddini Basavegowda of Keredimba Village—respected Solega elder, tammaḍi, and teacher.

Acknowledgements

Foremost in my thoughts are the Solega people who shared so much of their time, knowledge, and culture with me, warmly welcomed me into their homes, and patiently answered my endless questions with unflagging good humour. It has been a privilege to work closely with M. R. Madha, Heddini Basavegowda, Nanjegowda, Tammadi Dasegowda, Jadeswami, Madha, Jadeya, Kumbegowda, Ketegowda, and Javana, all of whom, with characteristic generosity of spirit, forgave my inability to become fluent in the Solega language, and to master the hundreds of plant names known to them. Dr. C. Madegowda, along with Nanjegowda, played a key role in identifying new Solega speakers for me to work with, and introducing me to the wider Solega community. Much of the work presented here would not have been possible without their enthusiastic support. The other staff members of the field station, where I spent many productive months, are also remembered with fond gratitude for making me feel at home—Renukamma, Nagendra, Gayatri, and Rajanna.

My supervisors, Professors Nicholas Evans, Andrew Pawley, and Alan Rumsey, each went beyond the call of duty in providing me with the benefit of their knowledge and experience, and in dragging me out of comfort zones that I had grown far too accustomed to. I am truly grateful not only for their intellectual input—needless to say, this book is greatly improved as a direct result of their guidance—but also for their friendship and kindness.

I am fortunate to have two wonderful collaborators in India, who allowed me to explore topics that I had little expertise in. Samira Agnihotri of the Indian Institute of Science, Bangalore, not only taught me everything I know about birds, but also helped me develop an appreciation for both scientific ornithology and Solega bird lore. Her friendship and company at the field site is also greatly appreciated. Professor Sankara Rao, also of the IISC, kindly offered his vast expertise in the taxonomy of the plants of southern India, to help with the labelling and identification of hundreds of plant photographs.

My early attempts to learn Kannada were serenely endured by Dr. Ananth Rao and Dr. Ajay Narendra, and I am grateful for their generous instruction in the intricacies of their mother tongue. Ajay was also the person who first suggested that I work with the Solega, and I thank him for his timely advice.

For providing many hours of editorial support during the last stages of manuscript preparation, I thank my friend Vincent Chea.

I thank the Australian National University and the Hans Rausing Endangered Languages Project for providing the funds that made possible three visits to the field site. I am also grateful to the Ashoka Trust for Research into Ecology and the Environment—particularly to Dr. Siddappa Setty and Dr. Nitin Rai—for allowing me to use ATREE's field station and other facilities on multiple occasions. The publication of this book was supported by a Faculty of Arts Publication Subsidy Scheme from the University of Melbourne.

Finally, I acknowledge the love and support of my parents, U. Aung Nyein and Daw Than Than Nu, whose numerous sacrifices have allowed me to follow my interests down many winding paths.

Contents

Chapter 1
Introduction

Fifty years in the past, a Solega person might have readily volunteered, "*Namma ka:ḍu senda:gade*", "Our forest is beautiful", when asked to describe their homelands in the Biligiri Rangaswamy Hills (B. R. Hills) of Karnataka State, southern India. This was a time when the forest understorey was dominated by tall grasses, and old trees fell only to be replaced by young saplings, when wild mammals, big and small, were plentiful and well fed, when the fragrant honey gathered by ground-dwelling bees from wildflowers provided sustenance to people on long forest walks, and when families were free to clear patches of the jungle with fire to grow crops. Today, Solega elders are far more likely to say, "*Namma ka:ḍu senda:gittu*" "Our forest *was* beautiful". The changes that have occurred to the ecology of the Solega's forests in the last five or six decades—in particular, invasion by the woody weed *Lantana camara*—have had a devastating effect on biodiversity, the behaviour of wild animals, and the regenerative capacity of the forest, according to local observers [1]. Moreover, they have radically altered human interactions with the forest, as common foods, medicines and building materials become scarce, and ancient forest trails and sacred sites risk disappearing under impenetrable thickets.

This book attempts to present Solega ethnobiological knowledge as a coherent system that has survived these changes for the time being. Both the Solega language and ethnobiological knowledge are currently threatened by language shift (to the locally dominant language Kannada) and lifestyle change, partly because of institutional pressures, and partly as a result of increased contact with mainstream Indian society. There is, consequently, an urgent need to document not only the language in its own right, but also those elements of cultural heritage that are encoded in language. The following chapters are a documentation of Solega ways of conceptualising the forest, its organisms and associated natural phenomena. It is centered primarily on the ideas and relationships encoded in the Solega language, as well as the encyclopaedic knowledge of the people who speak it. A variety of topics is investigated in these chapters, ranging from ethno-classification to detailed life-history descriptions of a single group of culturally-important organisms.

A. Si, *The Traditional Ecological Knowledge of the Solega*, Ethnobiology,
DOI 10.1007/978-3-319-24681-9_1

1.1 A Brief Introduction to the Field

The field of ethnobiology has come far in a short period of time, and currently engages practitioners from a range of academic backgrounds. In his review of the state of ethnobiology at the turn of the millenium, Ford [2] listed as many as 16 different sub-fields in which papers had been published in the *Journal of Ethnobiology* since its launch in 1981. These covered subjects as diverse as classification, conservation, nutrition, pharmacology and zooarchaeology, to name just a few. Papers dealing with linguistic issues formed a distinct minority in Ford's tally, and nearly all of these focused on ethnoclassification as their object of study. In contrast, this book takes a holistic, but consciously language-centred, look at the knowledge people have of the natural world, and investigates topics that include not only folk classification, but also folk ecology at the level of the landscape, semiotic knowledge in terms of meaningful signs and the relations between plants and animals (including humans), and the detailed knowledge of the life history of a particular group of organisms. First however, I provide a brief introduction to the field of ethnobiology and its connections to human language.

1.1.1 Documenting Language and Traditional Knowledge Simultaneously

The ethnobiological knowledge or traditional ecological knowledge (TEK) of non-industrialised societies has, in recent decades, come to be viewed not only as an important part of a community's cultural heritage, but also as a vital resource for researchers involved in activities like conservation biology [3, 4]. As a result, many biologists are now calling for an active engagement with such communities, with a view to making them stakeholders in any conservation efforts. In many cases, such partnerships have led to real-world conservation outcomes that have benefitted both the community and the natural environment [5, 6]. There is also a growing awareness that the work of linguists and anthropologists only further strengthens this enterprise, as their culturally-sensitive 'emic' perspectives perfectly complement the biologists' 'etic' compendium of objective facts [7]. The work of language documentation is similar in many ways to that of conservation biology, in that both are contingent on a strong appreciation of diversity. While it is heartening that language documentation has developed into an independent field of research in recent years, this young discipline also has much to gain by engaging with other, complementary fields. Speakers of small, endangered languages, especially those situated far from urban centres, routinely engage with their natural environment, as they go about the mundane tasks of obtaining food, fuel, water and building material. The languages of such communities come to encode much encyclopedic knowledge about biological and ecological entities and phenomena. This knowledge is as important as the knowledge of religious practices, local customs and taboos and kin-based relationships in allowing a person to be a fully-functioning member of his/her community.

1.1.2 *Defining 'Ethnobiology'*

'Ethnobiology' is practiced in many guises by researchers with diverse skill sets and academic persuasions, and so it is unsurprising that this term now encompasses studies that approach the investigation of TEK from a variety of angles. The following section, which describes some of the main focus areas of ethnobiological research, contains extracts from a paper published in the journal *Language Documentation and Conservation* [8].

An obvious entry point into the biological domain in a given language is the naming and folk taxonomy (classification) of living organisms in that language. This facet of ethnobiological knowledge has generated much research interest in recent years, with considerable effort being expended on the question of whether there are universal patterns in folk classifications across the world's languages. An influential publication in this respect was Brent Berlin's [9] *Ethnobiological Classification*, a summary of more than two decades of research by Berlin and his colleagues on this topic (see also [10–12]). This monograph presented evidence from unrelated languages to make a case for the existence of many linguistic universals in folk classification and nomenclature. Since then, there has been a flurry of reports from ethnobiologists scattered around the globe purporting to 'confirm' the claims made in Berlin. Some researchers remain sceptical, however (e.g. [13, 14]), and maintain that far more languages need to be investigated in detail in order to address the issue of universals.

A language community's knowledge of the natural world cannot be easily teased apart from what might loosely be termed 'cultural' knowledge; the latter, in turn, often runs seamlessly into the domain of religious belief. The interaction between the seemingly objective knowledge of the natural world and a community's subjective cultural attitudes and belief systems can be an interesting field of study in itself. This is best exemplified by the writings of the English anthropologist Ralph Bulmer, who worked extensively with the Kalam people of Papua New Guinea in the 1960s and 1970s. In his classic paper *Why is the cassowary not a bird?* he convincingly demonstrates there are many reasons why Kalam speakers do not classify this large flightless bird as a *yakt* 'flying bird or bat'—these reasons go beyond mere objective 'facts' such as physical appearance or lack of flight, and are instead firmly situated in the very special (kin) relations that cassowaries are meant to share with humans in Kalam mythology [15]. Incidentally, Bulmer [16] himself suggested a typology of ethnobiologists, including investigators whose primary orientations were: (1) lexicographic, with an emphasis on biological vocabulary, (2) formal, in that they focussed on taxonomic logic, (3) social, Roy Ellen being prominent among these (see below) (4) biological, i.e. professional biologists who develop an interest in traditional knowledge systems, and (5) natural–historical, of which category he claimed membership.

The management and use of a particular natural resource by a community has frequently been investigated by anthropologists and ethnobiologists. A good example in this respect is the study by Roy Ellen on the cultivation and harvest of sago palms for their edible starchy pith by the Nuaulu of eastern Indonesia. Ellen [17] discusses how a variety of factors—ecological, genetic and anthropogenic—have conspired to minimise the visible morphological variation in sago palms, and how this has resulted in a reduced number of varietal names for this species (in contrast to other heavily cultivated species like rice and bananas) in various languages in this part of the world.

Indigenous knowledge of local ecosystems, and the species contained therein, has great potential to inform scientists and conservationists of hitherto unknown aspects of ecology and behaviour. It has been noted that the 'diachronic' knowledge base (diachronic, in that the knowledge has been collated over innumerable generations) of traditional peoples, perfectly complements the 'synchronic' observations of western science [18]. The past decade has seen a great deal of interest in such 'applied' aspects of ethnobiology, as scientists come to realise that much can be learnt, from indigenous peoples, about sustainability, natural resource management or even basic biology [3]. Subsistence or artisanal fishermen, for instance, can be a valuable source of information of the breeding habits, diet and migration patterns of commercially-important fish (e.g. [19]). The information gathered in such studies is often new to science, and may prove crucial to the management of rare or endangered species.

The types of studies described above have the potential to document information about the natural world that is valuable not only to the indigenous communities consulted in the studies, but also to academics in a range of fields. However, one major drawback of some ethnobiological studies is that the data are presented in the academic literature in a form that is, for all practical purposes, inaccessible to the community.

Such studies not only inadvertently deny their consultants the ability to access their own traditional knowledge, but also implicitly value 'content' over 'form'—that is to say, the biological information, over the language that encodes that information. In recent years, however, there has been a significant emphasis in ethnobiological research on the return of TEK to local communities in a usable form. Notable examples include projects such as the People and Plants initiative jointly funded by the Royal Botanic Gardens in Kew, UNESCO and WWF; the Terralingua project, which seeks to promote biocultural diversity; and also individual researchers who aim to produce multilingual resources such as the Tok Pisin and English *Reite Plants* handbook [20].

Modern ethnobiological textbooks and field guides now regularly include a section on language—for instance, the useful introduction to basic linguistic concepts and methodologies in Gary Martin's *Ethnobotany* [21]. Similarly, a section in Fikret Berkes' *Sacred Ecology* draws the ethnobiologist's attention towards various linguistic issues that could confuse the task of data collection, and also warns against clinging on to one's own personal linguistic prejudices while conducting fieldwork [22]. More recent edited volumes such as Maffi [23] and Anderson et al. [24] contain comprehensive listings of articles that describe current efforts across the globe to protect and foster both linguistic and biological diversity.

Among field linguists, serious engagement with biological phenomena has been slow to manifest itself, although there are notable exceptions to this trend, such as Julie Waddy, Nicholas Evans and Murray Garde [25–27]. A reluctance to combine the documentation of TEK and language appears all the more puzzling when one considers the many inter-disciplinary enterprises that have become commonplace in the traditional linguistic research agenda—investigations into cultural domains such as kinship, physical domains such as landscape and psychological/cognitive domains such as colour terminology and spatial representation, just to name a few. The new, but rapidly growing field of language documentation puts a great premium on the obtaining samples of different, culturally relevant speech genres [28], and I have argued [8] that documenting TEK is an ideal way in which to achieve this. Much has been written on the negative consequences of language death, and possible solutions to counteract it [29–31], as well as on the practice and method of language documentation [32, 33]. Of these, only the volume by Nettle and Romaine (briefly in a section on 'Indigenous Knowledge Systems, p. 166–167) and a chapter in Thieberger's *Handbook* [34] explicitly mention TEK. Traditional biological knowledge would arguably rank as one of the most important topics of conversation among members of non-industrialised communities; it could be claimed that kinship systems are talked about at least as much as biological phenomena in some societies, but it is hard to imagine a language community obsessed with discussing the colour of objects. Moreover, linguists are ideally placed to carry out ethnobiological research in collaboration with specialists in various fields of biology—the former, by virtue of attending to analyses of the formal grammatical features of language, and to the way meaning is created in context, have a better chance of avoiding the misunderstandings inherent in cross-cultural communication.

1.2 Language in Ethnobiology: A Classificatory Bias

'What is named, and how?' is one of the fundamental ontological questions of linguistics, and can be answered to a significant extent by a consideration of plant and animal names in a given language. Brent Berlin [9] has been highly influential in this respect, arguing in his monograph *Ethnobiological Classification* that there are predictable ways in which species get singled out for naming cross-linguistically, and also that there are regular patterns in the way these species are labeled (see Sects. 2.2.1 and 2.2.2 for a listing of the main points of Berlin's model). This is an important idea that clearly needs to be tested thoroughly via the collection and careful analysis of culturally-sensitive, contextualized ethnolinguistic data from a large number of unrelated languages. Berlin's ideas, developed over a period of around two decades, have proven so attractive and influential that the great majority of language-centric studies carried out in recent decades by ethnobiologists have been on the topic of folk classifications. Ford's [2] review of progress in ethnobiological research clearly shows that of the 218 articles published in the *Journal of Ethnobiology* since 1981, only a handful (26, or 12 %) could be said to discuss language-related issues, and of these, 19 focused entirely on folk classification. Although Berlin was careful to use phrases such as 'general principles' and 'empirical generalizations' while laying out the features of his theory (pp. 20–35), these 'general principles' have since come to be viewed largely as cross-linguistic universals, with Berlin himself embracing the use of the latter term [35]. Moreover, by insisting that ethno-classifications are "*largely immune from the variable cultural determinants found in other areas of human experience*" ([9], p. 9), Berlin has legitimized and propagated a view of language-based ethno-biological research, according to which it is perfectly acceptable—indeed, preferable—to restrict one's attention to the elucidation of taxonomic hierarchies of named organisms in a void bereft of most contextualising cultural references.

This is a great pity, for linguistic research on (non-classificatory) biological themes has time and again shown its potential to be of great interest and benefit to community members, and to workers in other professions (see [36], p. 20 for a review of some cross-disciplinary studies). Garde et al. [37] present, with extensive original language transcriptions, the Bininj Gunwok (Arnhem Land, Australia) people's knowledge of fire ecology and seasonal cycles, and their effect of living organisms. The inclusion of language material ensures that the knowledge and concepts of the speakers is transmitted with minimal alteration by the ethnographer or the analyst. At the same time, the fact that such research is seen to be a part of a co-operative effort between government research agencies, policy makers, natural scientists, linguists and indigenous peoples speaks volumes for the ability of language-based studies to document TEK, and disseminate it among multiple stakeholders.

The potential of language-based studies to uncover facts about the natural world that were, to that point, unknown to science has been repeatedly demonstrated—possibly the best known of these are the collaborative works by Saem Majnep and Ralph Bulmer [38, 39], which present fascinating accounts of the natural world from the point of view of the first author—a hunter, and speaker of Kalam, from the highlands of Papua New Guinea. The first contained a brief mention of the unpalatability and unpleasant consequences of eating the *wobob* bird (*Pitohui dichrous*) 15 years before it was reported by scientists as the first ever account of a bird with toxic feathers [40]. The second of these volumes provides frequent references to the Kalam term *abn* (glossed by Bulmer with the neologism 'undercroft'), which is an almost subterranean (in reality, it lies among the tree roots and decaying vegetable matter on the soil's surface) labyrinth of tunnels, which are home to small edible rodents—a world that was, until then, unknown to biologists.

Similarly, Evans [25] presents evidence from a range of Australian languages to demonstrate that polysemous flora–fauna terms can encode crucial ecological information that links together two species. This can include predation (a grasshopper and its preferred grass food have the same name), spatial collocation (a heron and the mangrove tree in which it nests), and temporal co-incidence (two species that regularly appear at the same part of the seasonal cycle). Linguistic data in the biological domain can also be used to reconstruct historical events [41] or even entire lost worlds (reviewed in Evans [36], Chap. 6). Notable examples in this regard include the outputs of the ongoing *Lexicon of Proto Oceanic* project [42, 43], which provide reconstructions of key plant and animal terms in an ancestral form of the Oceanic subgroup of the Austronesian language family, thereby providing a glimpse into the biological world of prehistoric humans from around 2000 BC.

1.3 Questions

In many ways, this book is a reaction to the research agenda set by the findings and predictions presented in Berlin, which, as noted in Ford [2], has led to a restriction of the scope of language-centred ethnobiological enquiry to names and classification schemes. The documentation of the indigenous names of locally-occurring plants and animals and the investigation of folk taxonomies is of course important, but as I have argued in [8], it is equally important to look beyond the lexicon, and to utilise the tools of field linguistics to uncover the great mass of encyclopaedic knowledge that native speakers associate with each item in their ethnobiological lexicon. The same can be said of the dense networks of relationships that people perceive as forming undeniable links between seemingly disparate named entities or phenomena.

1.3.1 The Ethnobiological Lexicon

I address issues of nomenclature and taxonomy early in the book, in order to focus on more holistic issues in later chapters. A basic, but important, question that relates to the lexicon is 'What is named?' while a logical second question might be 'How are named entities organised in a person's mind?' As is further discussed in Chaps. 2 and 3 of this book, Berlin places much emphasis on the perceptual properties of living organisms, predicting that those with striking morphology (bright colours, large size, or other features that make them 'stand out') will be the ones preferentially named by any language community.

Another key prediction made by Berlin is the universality, across human cultures, of ways of naming and classifying plants and animals. The reason for this, says Berlin, is that all humans possess the same cognitive capabilities, by virtue of which they should be able to detect the same form-based discontinuities in the biological world, and subsequently construct very similar classification schemes for living organisms. As evidence, he cites the experiment he carried out with American university students (further discussed in Chap. 2), who consistently arranged groups of unknown bird species in much the same way as the Huambisa and Aguaruna Jívaro people, on whose territories those birds were to be found.

Berlin's experimental results are certainly intriguing, but it is pertinent to ask what they really mean, and what conclusions may be safely drawn from them. A related claim is that when given a mixed set of names of plants and animals, subjects will invariably place the plants in one category, and the animals in a separate category, even if the subjects' language has no overt lexemes for 'plant' or 'animal'. This is meant to indicate the pre-existence of discrete 'plant' and 'animal' categories in the subjects' minds, which form the top nodes of the subjects' ethnobotanical and ethnozoological classification systems respectively. As an investigator approaching ethnobiology from a language-centric viewpoint, I would be interested to know whether a crucial element of Berlin's reasoning is supported by empirical linguistic data: that the categories that subjects are *able to* construct in experimental situations because of shared human cognitive facilities really do represent the categories that they would *normally* make use of in their everyday interactions with their external environment (including other humans). In other words, do the people, who lack a 'plant' lexeme in their language, but can still discriminate between (biological) plants and animals when asked to do so, routinely make reference to the category of objects denoted by, say the English 'plant'?[1] Similar objections have been raised by linguists dissatisfied with their field's disproportionate emphasis on grammaticality, at the expense of naturalness. For instance, Pawley and Syder [45] have argued that while a formal grammatical analysis goes a long way towards explaining what can be said in a language, a proper understanding of the phenomenon of nativelike selection (i.e. selecting only natural and idiomatic sentences from the infinite utterances that the grammar allows) can only be gained through an investigation of phenomena such as 'lexicalised sentence stems', which give an indication of what native speakers actually say.

Parallel to the issue of a 'perceptual' explanation for human classificatory behaviour is the question of the role of 'culture' in the demarcation of named categories. There has been a resurgence of interest, in recent years, in detecting the presence of Whorfian (relativistic) effects in a variety of languages and semantic domains. Many psycholinguistic studies have compared the languages of different speech communities or of multilingual individuals to show that particular languages do indeed impose certain constraints on human perception, with both chronic (long-term) as well as "online" (short-term) effects [46–48]. Retuning to classifications, a 'perceptual' categorisation might motivate a person to discriminate between metal and wooden chairs (i.e., create two temporary, *ad hoc* categories) in the context of, say, deciding which bits of furniture should be placed outdoors, but one might expect a 'cultural' categorisation to be more stable because, by definition, it would be a categorisation learnt either from one's parents, or through formal education, or through regular interactions with other members of one's community. A good example of such a categorisation might be the highly culturally-sensitive judgements of substances or organisms deemed 'edible' or 'inedible' (or, perhaps more appropriately, 'appetising' vs. unappetising'). Tripe, blood, sago grubs, fish paste and blue cheese can be either delicious or revolting, depending on the cultural milieu within which the investigation on edibility is carried out. It has been argued that a category that comprises 'edible' substances is formed very early in an infant's life [49], but can the same be said for categories that manifest themselves in experimental situations, but subjects do not have a linguistic label for? Berlin claims that utilitarian and other cultural factors account for very little of the structure and content of a folk classification, as the perceptual features of the organisms included in a classification are the main determinants of the ways in which they are categorised. In Chap. 4, I test this generalisation through a consideration of Solega bird names, and ask whether perception really does play a much more important role than culture.

[1] Another pertinent issue at this point is whether the semantic ranges of the biological term 'plant' and the vernacular 'plant' labels in different languages really do overlap, and if not, what the differences are (see [44], p. 315 for further discussion).

Another reason to be suspicious of the assumption, that a category that *can be* created is a category that *is normally* relevant in real-life situations, is the mass of evidence from controlled psychological investigations on category formation, learning and use. Much of the literature dealing with the flexibility and context-dependence of mental concepts dates to the 1980s and early 1990s, i.e., around the time when Berlin's *Ethnobiological Classification* was published. The psychologist Lawrence Barsalou has written a series of influential papers on the topic of flexible categories, arguing, for instance, that while "*different people* [in a speech community] *store very similar information for the same category in long-term memory...* [the] *tremendous flexibility that we have seen in... experiments arises not from differences in knowledge, but from differences in the* retrieval *of this knowledge* ([50], p. 34). This flexibility further manifests itself in the way people construct and use *ad hoc* categories comprising "*highly specialized and unusual sets of items*" ([51], p. 211) to meet short term goals, such as planning future activities. Such categories share some properties with 'common' (i.e. long-term) categories, but differ in that the former are not well established in memory, and show high inter-subject variability in the absence of a context. By way of explanation, Barsalou theorised that:

> Because ad hoc categories are so specialized, it may be optimal that perceiving an entity does not activate all the ad hoc categories to which it belongs. Seeing a chair and having categories such as "emergency firewood", "fits in the trunk of a car" and "used to prop doors open" come to mind would be highly distracting when these categories are irrelevant. Ad hoc categories should come to mind only when primed by current goals. (p. 223)

In a recent book chapter, Barsalou et al. [52] make a strong case for the inclusion of context in psychological research (as well as other domains of academic investigation). The authors point out that there is an overwhelming amount of evidence clearly demonstrating context effects on diverse phenomena, but more importantly, that taking context into account usually explains much of the variation present in data. Many theories claim that expert performance is more the result of simple pattern matching rather than reasoning, say the authors, and that the former is facilitated by storing situation-specific chunks or exemplars in long-term memory. Unfortunately, many psychological concepts are routinely tested and modelled in experimental situations where variation is ignored or treated as psychological noise, or where the context is strictly controlled to minimise variability [53]. In psychological studies on concepts and categories, in particular, there tends to be an assumption that categorization is primarily a bottom-up, stimulus-based process [54], whereas in reality, humans show variable categorisation behaviour depending, among other factors, on the situation or task at hand [51, 55], expertise [56] or language repertoire [57, 58].

A major implication of the preceding discussion for research on ethnobiological classifications is that the methods by which folk taxonomies are investigated (i.e. semi-structured interviews, sorting, grouping and identification tasks) may in fact represent but one type of context, within which one type of categorisation scheme can be obtained. Such a 'standard' Folk Taxonomy of X Group of Organisms in Language L carries with it the risk of not faithfully representing other, legitimate ways people may have of thinking about X, by virtue of having been elicited in a situation freed from the usual contexts that speakers of Language L might encounter on a regular basis. Indeed, the possibility that such taxonomies may well be artefacts of the analyst's mind has been raised by several authors [14, 59, 60]. I discuss the issue of methodology further in Sect. 1.4.

In their introduction to a study on the organization of food categories by English speakers, Ross and Murphy [61] made a similar observation: namely, that earlier psychological work on the classification of real-world concepts had "*often suffered from three limitations: a single hierarchy, a single function, and isolated knowledge*" (p. 496). This means that researchers often ignore the various cross-classifications that named entities may belong to, assume that classification is the only function for which concepts are used (while ignoring other functions such as induction, explanation, problem solving, category formation and communication), and focus on a certain kind of knowledge in isolation from much of the other knowledge that humans possess. Ross and Murphy presented their subjects with a long list of foods, and asked them to generate the categories that those foods belonged to. The researchers found that subjects were just as likely to create 'script categories' (categories that usually made reference to the time or situation when the food was eaten, or to the healthiness of the food), as they were to name standard 'taxonomic' categories, such as breads, meats, etc. The authors distinguished between script categories from Barsalou's ad hoc categories which, in the context of food, might include 'foods that are often cooked in water' or 'foods that squash easily'. Next, subjects were divided into three groups, and asked to sort the same food categories according to (a) taxonomic groupings, (b) script groupings, and (c) and any criteria the subjects found appropriate. Groups (a) and (b) produced groupings in line with their instructions, but although group (c) produced predominantly taxonomic groupings (56 %), a significant proportion (30 %) of their groupings were still along the lines of script categories (e.g. junk foods, breakfast foods). Interestingly, even in group (a), 22 % of the groupings were script categories, in spite of strict instructions to the contrary. Finally, the authors found that both script and ad hoc categories showed large priming effects, in that the presence of contextual information could motivate subjects to place items into these categories. Script categories could be spontaneously activated by the presentation of a food item, although this activation was not as strong or as consistent as that of the taxonomic categories.

How might these results relate to folk biological classifications? First, they show that it is quite normal for people to have more than one way of categorizing objects that they regularly interact with. Such alternative categories may exist long-term, and therefore be as perceptually salient as the more conventional, taxonomic categories. More importantly, they suggest caution while positing 'covert categories' (folk taxa that are not named, but that are often grouped together in sorting tasks; Berlin, 1992, pp. 139–160) as legitimate nodes in folk classifications. Naturally, such taxa would be valid if speakers were to consistently, and spontaneously, say that certain organisms 'belong together', or if there existed certain complex expressions in the language that made reference to an unnamed category. In the absence of such supporting information, however, it would be reasonable to regard covert categories with suspicion, as they could well be equivalent to the script categories or ad hoc categories described earlier.[2] This applies not only to covert groupings of 'folk generics' (which Berlin labels 'intermediate taxa'), but also to 'kingdom' level groupings, which are usually unnamed across languages. Ross and Murphy (1999) provide an illuminating example to demonstrate their point that highly specific contexts may motivate the construction of novel categories:

> For example, one may not have a well-established category of foods eaten at the movies, but one can easily construct such a category post hoc, including popcorn, soda, certain candies, and ice cream. If one often eats at the movies, this information may become more and more saliently represented for these items, until it can be as important a way of representing them as their taxonomic categories. (p. 540)

'Often eating (e.g. popcorn) at the movies' is reminiscent of the 'activity signatures' that Hunn [62] suggested as being of value in gauging the utilitarian significance of a particular plant or animal. Unfortunately, although covert categories such as 'doves' or 'birds of prey' [63] are frequently allowed on the basis that certain names tended to clump together in free recall lists, researchers seem to be dismissive of utilitarian groupings of species offered by speakers, such as 'plants that bear edible greens', or 'birds with (useful) ornamental feathers'. Berlin argues that such culturally-based covert categories are rare, and that they *might be better described as part of a cross-cutting system of classification*" (p. 152). Priming is a relevant phenomenon in such cases, as speakers may first categorise those taxa which can be grouped primarily by their morphological characteristics, and continue to use the same criterion to create subsequent groupings on an ad hoc basis. As a result, groupings based on utilitarian factors may be under-represented or missed altogether. The context of an elicitation session is also a highly unnatural one, in contrast to speakers' normal interactions with the plants and animals that the ethnobiologist wishes them to categorise. In the absence of the usual contextual cues that would normally accompany the categories being focussed on, speakers could effortlessly and unconsciously resort to whatever cues do remain in the task at hand. These would invariably be morphological cues, and could just as easily be the only cues available to a speaker in a name or specimen sorting task, or a free-listing task.

For such reasons, I have tried to avoid presenting formal Solega folk taxonomies of any group of organisms in this book, unless such a taxonomy was strictly necessary to make a point. Instead, I investigate some key assumptions that underlie Berlin's general principles, chief among these being certain misconceptions about the nature and practice of biological classification, as carried out by professional taxonomists. In Chaps. 2, 3 and 4, I also examine many other claims made in Berlin [9], including those relating to nomenclature, and to the different levels of the hierarchy of a given ethno-classification system.

---

[2] Ross and Murphy [61] did not present strict criteria to distinguish between these two types of categories, and it is possible that the difference is simply a matter of frequency of usage, with ad hoc categories only being used in very limited contexts. Another way of stating this would be that certain, very specific, kinds of contextual information are required before an ad hoc category is activated.

One way to ensure that folk taxa are described in ways that are relevant to the speech community is to ensure that the context of classification is not ignored. In Chap. 2 for instance, I argue that the classification of culturally important organisms such as honeybees and mushrooms only makes sense in the context of their patterns of use, while in Chap. 4, I show that the (socio)linguistic context has a bearing on the way birds are named, both in structured tasks, as well as in spontaneous discourse. Yet another way to ensure that culturally-significant and linguistically-relevant information is not disregarded is to make 'context' itself the object of study. Many themes discussed in this book, especially in the later chapters, have little overt connection with folk classification, as it seemed not only interesting, but also important, to investigate the encyclopaedic knowledge that Solega speakers associated with each named organism or natural phenomenon, as well as the knowledge of the relationships linking these organisms and phenomena. In doing so, I hope to move away from a taxonomy-centric paradigm of linguistic ethnobiology, and give traditional ecological *knowledge* the recognition it deserves to get from those who would study human language.

1.3.2 Analysing One 'Context'

The possibility that the context of an ethnobiologist's investigations (e.g. the experimental task) could result in the formation of short-term ad hoc categories has already been discussed above. This would be an example of an artificially imposed context leading to the creation of possibly artificial (from the speech community' point of view) folk taxa. It seems reasonable to assume, then, that a sound appreciation of the complexities of a community's TEK can only be gained by also studying the long-term contexts within which the knowledge is embedded, learnt and used. Such contexts are numerous and inter-related, and might include the community's geographical location (and accompanying biodiversity), cultural institutions and practices, linguistic history, migration history and history of contact with other communities, to name just a few. The study of each of these topics is deserving of an entire research project, and is understandably outside the scope of this book.

One context-providing topic that I did manage to investigate during my field research, and that arguably has the most direct bearing on TEK, is the Solega conception of the numerous landscape and forest types for which they have names. The physical landscape and the ecosystems, within which various named plants and animals are to be found, are likely to have a significant impact on Solega conceptions of those organisms. Accordingly, in Chap. 5, I investigate the features that Solega speakers attribute to each landscape or forest type, in order to determine exactly how plants and animals are linked to their environment.

I mentioned earlier the very likely prospect that Solega people view plants and animals not as isolated species, but as nodes that support a dense web of ecological interactions. Might this way of perceiving named entities also exist on a much larger scale, namely that of the entire landscape? In other words, is it possible that named ecosystems (i.e., forest or landscape types) are also linked, in the minds of the Solega, by a network of processes and interactions? The primary data on which the investigations of Chap. 5 are based are not completely naturalistic, as they were elicited through interviews. This is not a problem, however, as long as the limitations of data gathered under controlled conditions are fully acknowledged. In this case, the major limitation is that people would be prompted to explicitly provide information that would otherwise be completely implicit in everyday social discourse. Again, this does not invalidate my study, because the primary aim is to ask "What do Solega people know about landscape X?"—here, it is the Solega speaker's corpus of implicit encyclopaedic knowledge that is the object of inquiry.

In contrast, caution would be well-advised in the case of *how* questions, such as "How do Solega people perceive the interconnectedness of different ecosystems?" I do ask such a question later in the chapter (summarised in the following paragraphs), but this question is partly answered by means of supplementary information gained from unstructured Solega narratives, where the speakers were free to choose the direction in which the narratives progressed. The theoretical limits of a Solega conceptual system (here, the 'cognitive map') are therefore first established through the data gained from interviews, while the more spontaneous data provide glimpses into how the system is actually utilised in socially acceptable ways in everyday discourse.

The concept of a 'mental map' or 'cognitive map' has been thoroughly investigated by a wide range of professionals, including neuroscientists, psychologists, investigators of artificial intelligence, cartographers and city planners. In spite of this attention, the very definition of a cognitive map, as well as the ways in which such a 'map' might represent knowledge of the external world, remains contentious, possibly as a direct result of the great variety of theoretical orientations among those who investigate such phenomena.

The question of which viewpoint is psychologically more dominant still remains largely unresolved, and Kitchin [64] proposes that the term 'cognitive map' should be used in a utilitarian way to "*represent the knowledge of, and interactions with, the everyday environment, and geographical information gained through other secondary sources such as* [man-made] *maps*" (p. 5). While describing the semantic ranges of the various Solega forest and landscape terms, I periodically address the question of what Solega people know about the components of these places, and how they interact with them. Here, it also seemed relevant to ask, "What is the nature of the Solega cognitive map?" and "How does the cognitive map interact with the Solega's encyclopaedic knowledge (see Sect. 1.3.4 below) of the entities and phenomena situated within the landscape?"

1.3.3 Incorporating Variation

I mentioned earlier the criticism levelled by Barrett et al. [53] at psychological experiments that either ignore variation, or seek to minimise it. Normal language use is also characterised by "*variability of a structured and regular kind*" ([65], p. 340), and this variability can take the form of (at least) phonological, syntactical and lexical differences between individuals [66]. It would be unsurprising not to find variation in folk taxonomies or TEK either between individuals or sub-communities of a language group; indeed, there have been notable ethnobiological studies that have primarily investigated variation in TEK as a function of different social variables [59, 60, 67, 68]. It is precisely because of the existence of (often widespread) variation that a folk taxonomy, which only illustrates one way of categorising organisms, should be regarded as an idealised abstraction, rather than a representation of how people really think. As Labov [66] points out:

> The existence of variation and heterogenous structures in the speech communities investigated is certainly well-established in fact… Each investigator feels that his own community has been corrupted from this normal [i.e. variation-free] model in some way—by contact with other languages, by the effects of education and pressure of the standard language, or by taboos and the admixture of specialized dialects or jargons. But we have come to the realization in recent years that this is the normal situation—that heterogeneity is not only common, it is the result of basic linguistic factors. (p.203)

In this book, I have made a conscious effort to take note of inter-individual or inter-community variation in Chaps. 4 and 5. In Chap. 4, I ask whether certain concordances between my data and Berlin's predictions regarding nomenclature at the 'folk generic' level of classification truly do reflect a pattern that pervades the entire speech community, or whether there is significant variation in the choice of lexeme(s). A baseline level of dialectal (phonological) variation is, of course, to be expected, given that the Solega live in several villages that can be several kilometres away from one another. My interest lay in documenting variation at the lexical level, where certain organisms might be known by completely different names by different individuals, or where the combinatorial patterns of a given lexeme (for instance, in the formation of compounds) might diverge.

1.3.4 'Encyclopaedic Knowledge' as an Object of Study

There have been calls from linguists for some time to extend the boundaries of linguistic research, in order to incorporate phenomena that were once regarded as extra-linguistic (e.g. [69]). Supporters of Cognitive Grammar have been influential in this respect, as evidenced by their incorporation of the notion of 'encyclopedic knowledge' into their theory. Arguing that "*no clear and nonarbitrary dividing line can be drawn between linguistic knowledge and world knowledge*" [70], cognitive grammarians advocate semantic analyses which fully acknowledge the role of pragmatics in shaping the meaning of utterances ranging from words to complete sentences. Similarly, among ethnobiologists, Ellen [71] has argued "*that all classifications are discursive practices situated in a given social matrix and general configuration of knowledge and ideas… and that they are products of specific histories.*" The study of the Solega "*social matrix*" lies well outside the scope of this book, but in Chap. 6, I have made an effort to elucidate some of the main features of the "*general configuration of knowledge and ideas*" within which named Solega plants and animals are embedded. Using semi-naturalistic speech data (semi-naturalistic, in that the data were obtained through interviews), I investigate the explicit knowledge that Solega people have of the forest and its inner workings, and of the ecological relationships between different named organisms. I also ask how Solega people regard themselves as fitting into this network of relationships, and why they should know so much about interactions between organisms that seemingly have no bearing on their own lives.

Chapter 7 takes the study of encyclopedic knowledge to its logical conclusion, by attempting to describe all facets of the Solega's knowledge of a culturally significant group of creatures—the honeybees. A key reason for undertaking this study was to demonstrate that a speech community can indeed possess highly detailed and sophisticated knowledge of biological entities and processes, even if their folk classification might seem simple or shallow in comparison to those of others. Berlin [9] has argued that there are fundamental perceptual differences in the ways which hunter gatherers and agriculturalists categorise the natural world. This dichotomy has already been criticised by Ellen [72], partly because there are few, if any, instances of people who are purely 'agriculturalist' or purely 'hunter-gatherer'. In the case of the Solega, small-scale agriculture has definitely been practiced for several generations, but these people have never been beekeepers. Being, in effect, honey 'hunters', they provide an excellent case study for a comparison with a pre-industrial beekeeping society. Greece at the time of Aristotle was indeed such a society, and the philosopher has kindly left us extensive writings on the honeybee-related lore of his beekeeping contemporaries. Comparing Solega honeybee TEK with that of Aristotle's Greece, I ask whether there are significant differences between the two, and if the latter system of knowledge is any better than the former.

1.4 Concerns About Scope and Method

As early as the 1970s, William Labov had voiced concerns about traditional linguistic research techniques such as textual analysis, elicitation and intuitions having little to offer by way of explaining the functioning of language as "*a vehicle of communication in everyday life*" ([73]; p. 108). Suggesting the observation of speakers as the most difficult—but probably the most rewarding—data collection method, Labov identified three reasons why linguists might prefer alternative procedures:

> One widely propagated belief which is used to discourage the study of ordinary language is that speech is incoherent. Chomsky has often remarked that the child must discard the largest part of what he hears as ungrammatical (1965: 58). This view is a myth based upon no evidence at all, except perhaps a few transcripts of learned conferences…
>
> Secondly, we find that most investigators describe their own community as exceptional, rife with dialect mixture and chaotic variation as compared to the homogeneous nature of traditional speech communities. But such homogeneous communities are also myths…
>
> We find a third ideological barrier in the claim that all such data belong to some other far-away discipline called the study of performance, to be realized when we have mastered the facts of competence. The distinction between competence and performance may have its uses, but as it is now drawn it is almost incoherent. (p. 109–110)

These barriers are probably still as valid for linguistic research today as they were in 1972, but linguists are increasingly turning to the study of language as a means of "*communication in everyday life*"—the burgeoning fields of enquiry into bi-/multilingualism, code switching, discourse/conversation analysis, formulaic speech and many others clearly demonstrate this trend.

Still, many commentators have argued for a further widening of the scope of linguistic investigation. Viewing grammatical and semantic complexity through the lens of kinship systems, Evans [74] puzzles over the reasons why the effects of culture on language structure tend to be excluded even in functionalist approaches to grammar. By his account, culture can select for language structure in two ways: (1) things that are talked about more by a language community (e.g. culturally salient phenomena such as kinship) are more likely to get grammaticalised; and (2) conversational implicatures that draw on mutually shared knowledge (might this overlap significantly with culture-specific 'encyclopedic knowledge'?) become 'semanticized', i.e., freed from particular contexts. While optimistic about the possibility that many instances of culturally-driven grammatical structure remain to be uncovered, Evans urges a disciplined, empirically-based and explicit methodology, like the one suggested by Simpson [75]. Ultimately, however, what is required is a radical overhaul of the way linguists go about the business of data collection:

> More complete accounts will also require linguistic practitioners to gather information on "not yet grammaticalized" collocations. The grammar-writing traditions of descriptive linguistics focus on structures that have already emerged; but to explain where they come from, we must look at the messier data of actual speech. This needs larger, more finely transcribed corpora. (p. 28)

Similarly, Linell [76] has criticized the research agendas of many linguists as 'monologic', and as a counter, has called for "*an emphasis on interaction and contexts, on language being used in situational and socio-cultural contexts, in interaction with others and with our physical and social surroundings*" (p. 157). A 'dialogic linguistics' is needed, says Linell, that engages with holistic chunks of linguistic praxis, such as communicative projects, episodes in discourse, communicative activity types, and 'social representations'. In studying these concrete manifestations of linguistic competence, a linguist must keep in mind that the meaning of utterances is more than simply the meaning of its constituent parts, and that interpretations emerge in different ways in different dialogical situations and communicative genres.

The preceding discussion has a great bearing on the kinds of research that could be carried out in ethnobiology as well. The situation here is not irretrievably bleak, and there are indeed ethnobiological studies that focus on language as the object of investigation. However, the abstract and the acontextual appear to dominate here as well, for most of these studies deal with the elucidation of folk taxonomies of living organisms. Notable exceptions, where naturalistic speech is presented as evidence to back up the author's taxonomic claims include Baker [13] and McKnight [77]. Unfortunately, a significant majority of taxonomic studies appear to be aimed at testing the universalizing claims made by Berlin [9], and tend not to venture beyond considerations of how named plants and animals can be hierarchically arranged in a given language.

The inclusion of naturalistic speech is essential when asking "*how*" questions in ethnobiological research (such as "*How do people normally group together named organisms?*"), as standard elicitation practices—interviews, naming tasks, sorting tasks—create artificial contexts that have the potential to generate unnatural responses. The idea that context and basic conversational principles can significantly influence the interpretation of an utterance has been known to linguists for quite some time [78]. Unfortunately, some ethnobiologists who work on folk classifications take their consultants' responses during elicitation sessions far too literally, ignoring the fact that many features of such responses would be affected by the very specific contexts of those sessions. Berlin, for instance, has argued that 'generic' level categories have a special status in folk taxonomies, because they are the ones generally provided in response to a question regarding an organism's identity. As an example, Berlin presents us with the following hypothetical conversation between an ethnographer and a consultant (adapted from p. 52):

Ethnographer:	That (over there), what's its name?
Consultant:	(It's named a) tree.
Ethnographer:	That (over there), what's its name?
Consultant:	(It's) also (called a) tree.

The above exchange is characterized by Berlin as 'improbable', 'ludicrous' (and rightly so), and is meant to support his argument that 'generic' taxa—the "*names of common speech*" (p. 53), the types of labels that would seem most appropriate in this context—would be used preferentially because they label the "*smallest fundamental biological discontinuities easily recognized in any particular habitat*" ('tree' in this case would belong to Berlin's 'life form' rank). I disagree with this conclusion, for Berlin is confounding cognitive salience with some other highly relevant features of "*common speech*", namely Grice's [78] maxims of quality (i.e. accuracy), quantity (adequacy of information), relevance (making utterances relevant to the issues currently being discussed) and manner (avoiding ambiguity). First, we need to assume that the ethnographer and the (willing) consultant are both desirous of obeying the Cooperative Principle.[3] It can then be safely assumed that since the ethnographer is a naïve (in the psychological sense) outsider, and since the consultant is aware of this, the latter party will try to accommodate the former's lack of knowledge by being maximally helpful and relevant. The consultant will have made the assumption that the ethnographer is probably familiar with trees, and because s/he is new to the field site, may choose not to burden him/her with a highly precise subordinate label. The only option remaining to the consultant—one that is simultaneously accurate, adequate, relevant and clear—is to provide a 'generic' label as a compromise. One can easily imagine specific scenarios where taxa of ranks other than 'generic' constitute the most appropriate response to an utterance, in keeping with the Cooperative Principle:

[3] The Cooperative Principle [78] states: "Make your contribution such as it is required, at the stage at which it occurs, by the accepted purpose or direction of the talk exchange in which you are engaged."

Q1: What's that?
A1: That's a Greater Bamboo Lemur. ('specific'/'varietal')

(A biologist leading a group of nature lovers on a wildlife trip)

Q2: What's that?
A2: That's a lemur. ('generic')

(A biologist addressing a passing tourist)

Q3: What's that?
A3: That's an animal. ('kingdom'[4])

(A biologist who has been identifying bird calls for a group of birdwatchers, and is asked to identify a novel call)

The above examples show that context plays a highly important role in allowing the respondent (the biologist) to decide on the level of specificity required to answer any given question. The relevance of Grice's maxims to the interview scenario is far from being a conclusive refutation of Berlin's claims about the primacy of 'generic' taxa. However, the maxims do demonstrate that elicited data are highly context-specific, and that naturalistic speech should be given pride of place in any ethnobiological investigation.

A system of TEK that is understood only through a classificatory system—elicited outside the normal contexts in which people interact with living organisms—is akin to a biological species that is perceived exclusively through the four-letter alphabet of its genome. It is true that an organism would not exist without its genome, but it is equally true that there is far more to an organism than a mere sequence of nucleotides. A genome predisposes an organism to certain structural, physiological and behavioural models, but the vast literature on epigenetic phenomena has demonstrated without a doubt that life is all about the interaction of genomes with the environments within which organisms grow, swim, crawl or fly. Similarly, an ethnotaxonomy serves a very limited purpose unless it is situated in the everyday contexts that the speakers of that language would recognize. A study in language-centered ethnobiology is incomplete unless it addresses the question of how people communicate with each other about biological themes; as argued above, such communicative performance is inextricably tied not only to norms of linguistic practice that are accepted by the community as natural, but also to the rich store of encyclopaedic knowledge that speakers possess of the entities and phenomena being discussed.

1.5 Book Outline

In the following chapters, I have attempted to capture not only the 'encyclopaedic knowledge', but also the "*situational and socio-cultural contexts*" that some of the authors cited above have deemed necessary, in order to properly understand the

[4] 'Animal' is a 'kingdom' level taxon according to Berlin [9], but Wierzbicka [44] has challenged this claim.

Solega language as a "*a vehicle of communication in everyday life*". Such an approach, coupled with the presentation of transcribed and translated utterances from my corpus of audio recordings, has allowed me to maximise the volume of semi-naturalistic (i.e. obtained from interviews), culturally relevant and contextualised linguistic data on which subsequent analyses and conclusions are based.

Briefly, the themes covered in this book begin with a consideration of Berlin's universals of nomenclature and classification (Chaps. 2, 3 and 4), a discussion of what is named, and why, and an illustration of dialectal variation in the naming of Solega birds (Chap. 4). Chapter 2 deals with some important theoretical issues that have a great bearing on how people construct and interact with such classifications. This chapter critically examines some key assumptions that are prevalent in the ethnobiological literature, and asks whether there really are 'universals' in the way people identify, categorize and name the living organisms in their environment. The plants of the Solega's forests are discussed in Chap. 3, with the focus being on nomenclature and its relationship to scientific taxonomy. These issues are contextualised through cultural notes on Solega people's attitudes towards plants, the central role that plants play in the coining of place names, the importance of large, individual named trees in Solega culture, and the myriad interactions that people have with plants of all sorts in their everyday lives. Chapter 4 continues the themes of Chap. 2, and demonstrates how real language data can often come into conflict with homogenising theories by describing, in detail, the naming and classification of birds in the Solega language. Here, the substantial discussion on classification and taxonomy is followed by an account of the very important role some birds play in Solega religion, folklore and everyday life—a second foray, following the ethnobotanical discussion of Chap. 3, into the world of encyclopaedic ecological knowledge.

The latter half of the book steps away from the question of taxonomies, and definitively enters the realm of the encyclopaedic and ontological. I begin with the largest possible scale—that of entire ecosystems. In Chap. 5, I attempt an informal semantic description of landscape and forest terms in Solega, and a discussion of possible relations between information structure and perceptual salience. The descriptions presented here demonstrate how the Solega perceive their forest home as an interconnected mosaic of habitat types. These habitats are home to distinctive assemblages of plants and animals which show seasonal life-cycle or migrational patterns—the Solega conception of these patterns is overlaid on their mental representation of the forest habitats, resulting in a temporally-updating map of natural resources (and other salient phenomena) in the physical–biological environment. Continuing the theme of large-scale processes and patterns, Chap. 6 includes a description of the natural cycles that the Solega routinely observe in their forest, and the ways in which these are encoded in everyday language. This includes not only the plant phenological cycles mentioned in the preceding chapter, but also the annual monsoon cycle, which is divided into several meaningful phases. I go on to describe smaller-scale patterns that Solega people often talk about, such as the spatial and temporal signs that indicate resource availability, and the relationships that other organisms have with each other. I then suggest a hypothesis to explain why the knowledge of such relationships should be important to the Solega. In Chap. 7, I zoom in from the macro- to the micro-scale, and describe in detail Solega knowledge of, and interactions with, one highly significant natural resource that the forest provides: honeybees and their honey. The Solega recognise four major groupings of honeybee, corresponding to four biological species. Even though they have never been beekeepers, their knowledge of bee biology rivals that of societies that have engaged in beekeeping for centuries. This last chapter also draws from debates in biological evolution to ask whether it makes sense to talk about languages 'evolving' more 'complex' structures over time.

1.6 Field Situation

The Solega live in the Biligiri Rangaswamy Hills (henceforth B. R. Hills) south-east of Mysore and south-west of Bangalore, the capital of Karnataka state. The B. R. Hills are a minor branch of a large mountain chain, called the Western Ghats, that runs along much of the western coastline of peninsular India. The Western Ghats are recognised as a 'Biodiversity Hotspot'[5] by international organisations such as Conservation International and the World Wide Fund for Nature. I spent a total of around 11 months at this location, split over three field trips (November 2008 to May 2009; August to November 2010; May to July 2011). The visits were planned to fall on non-overlapping months of the year, so as to allow me to record the widest range of biological phenomena occurring in a single annual cycle. During each trip, I was based at a field station (11.98741 N, 77.14396 E) run by the Bangalore-based non-profit conservation organisation Ashoka Trust for Research into Ecology and the Environment (ATREE). The field station is easily accessible by bus from the nearby towns of Yelandur (40 min) and Kollegal (60 min), the district capital Chamrajnagar (60 min), and the major urban centres Mysore (3.5 h) and Bangalore (5.5 h).

The Biligiri Rangaswamy Temple Wildlife Sanctuary (henceforth BRT) was established at this site in 1972, and is part of a chain of protected areas called the Nilgiris Biosphere Reserve.[6] This Reserve straddles the confluence of the Western Ghats with its mirror-image counterpart, the Eastern Ghats. BRT lies at an elevation of 750 m to 1800 m above sea level, and covers an area of approximately 540 km^2. It encompasses a range of vegetation types, including moist evergreen rainforest, tropical deciduous forest, scrub forest, grassland and shola forest, the latter being unique to the Western Ghats [80]. The sanctuary is home to at least 27 species of mammals, over 100 species of butterflies [81] and around 250 species of birds [82, 83]. In addition, around 800 species of plants have been recorded here [84].

[5] Hotspots are areas featuring exceptional concentrations of endemic species and experiencing exceptional loss of habitat. To qualify as a hotspot, an area must contain at least 0.5 % or 1500 of the world's 300,000 plant species as endemics [79].

[6] Biosphere Reserves are places officially recognised under UNESCO's Man and Biosphere Programme. They are defined as "places that seek to reconcile conservation of biological and cultural diversity and economic and social development through partnerships between people and nature".

The name of the wildlife sanctuary derives from the presence of an ancient temple to the god Rangaswamy or Ranganatha (alternate names for Venkateshwara, or Vishnu) on top of a bare granite hilltop (*biḷi giri* means 'white hill'). While the Solega do not worship Biligiri Rangaswamy, they show him a certain amount of respect, and claim kinship with him, as told by one of their myths. BRT is also home to the Vivekananda Girijana Kalyan Kendra (VGKK; 'Vivekananda Mountain-Dweller Welfare Centre'), a non-profit organisation that was first set up by Dr. Sudarshan, a physician from Bangalore, in 1981. Over the years, the organisation has grown, and radically improved the living conditions of numerous Solega families, particularly those living in the settlements surrounding VGKK. It now provides a free school (which includes a free afternoon meal) and basic health services through its clinic, and also takes part in various community development programs in collaboration with ATREE. For his decades of work with the Solega community, Dr. Sudarshan was presented with the Right Livelihood Award in 1994. More recently, he was runner-up for the 2009 Inaugural BMJ (British Medical Journal) Group Lifetime Achievement Award [85].

Some researchers of ATREE, who have carried out either biological fieldwork or community development projects in the B. R. Hills, have been in contact with the Solega for a few decades. ATREE employs a number of young Solega men to work as field assistants and drivers for researchers visiting the field station. They were the first Solega that I came in contact with, and soon proved to be knowledgeable and enthusiastic consultants for my own work. These consultants frequently introduced me to other, often senior, members of the community, who they said were either well-versed in particular domains of folklore, or were more fluent in 'pure' Solega. In this manner, they greatly facilitated my efforts aimed at documenting both their language and their traditional knowledge.

Data collection was carried out through open-ended interviews, with Kannada as the contact language, or through focussed question-and-answer sessions (primarily to elicit plant and animal names) during forest walks or visits to villages.

1.7 Ethnographic Sketch

1.7.1 The Community: Name and Identity

It was initially difficult for me to find any published material on the Solega due the range of orthographic renditions used by various authors for the name of this group. Variants include 'Solaga', 'Sholagar', 'Sholaga' (this is the spelling used in Ethnologue; ISO code: sle), 'Sholiga' and 'Soliga/Soligaru'. The latter spelling is the one currently preferred by the Indian media and by Kannada speakers—this applies to both English and Kannada orthography (in the Kannada script, this would be rendered as ಸೊಲಿಗ '*soliga*'). I have used the spelling 'Solega' in this book, as I have been told quite unambiguously that '*we are called Solega.* Soliga—*that's what outsiders call us!*"

The umbrella term 'Solega' can actually be applied to two quite different ethnic groups, each with its own language and customs. The Solega who live in the B. R. Hills are the *Aidu Kula* 'five clan' Solega, and the bulk of their population lives within the borders of the state of Karnataka. There is another community, the *E:ḷu Kula* 'seven clan' Solega, who live primarily in the neighbouring state of Tamil Nadu to the south, and whose distribution overlaps slightly with that of their northern counterparts. Apart from the fact that the *E:ḷu Kula* Solega have a very different social structure, their language is also noticeably different, even at a first hearing. Just as the *Aidu Kula* Solega language is closely allied with Kannada, the speech of the *E:ḷu Kula* Solega has many affinities with Tamil. The two Solega communities do not intermarry, and appear to have little to do with each other, in spite of a lack of animosity. Henceforth, 'Solega' will be used to refer exclusively to the *Aidu Kula* 'five clan' Solega of the B. R. Hills.

The Solega inhabit a large area in the B. R. Hills and surrounding plains, and many settlements are either a few hours' or a few days' walk distant from each other. An average Solega person will have relatives—close and distant—living in several other villages, but may not see them except in case of an illness, a festival or a family celebration. Nevertheless, many people will regularly expend the effort and/or monetary cost of traveling long distances to attend important events occurring in distant villages. The Solega are aware of the existence of other tribal groups living in the Western Ghats, but state that they have historically had little interaction with them.

1.7.2 Previous Studies

Very little published material exists on the Solega language, culture or traditional knowledge. A thin volume called 'The Soliga of the Biligiri Rangana Hills' was published by the Archeological Society of India shortly after the creation of the BRT wildlife sanctuary [86]. This informative, but occasionally inaccurate, ethnographic sketch gives a useful account of Solega life in the 1970s. A 'Sketch Grammar of the Soliga language' was published by the Russian Dravidianist Kamil Zvelebil [87], who unfortunately had access to only one young 'Shōlega' speaker who was also bilingual in Tamil. The data presented in this paper differ greatly from my own experience of B. R. Hills Solega, and it is very likely that Zvelebil has in fact described a variant of Seven Clan Solega. Zvelebil reports that his consultant's phoneme inventory contained the vowels transcribed as ï, ë, ä and ö, and this matches my own passing impressions of Seven Clan Solega. Such phonemes occur in the speech of only a few B. R. Hills Solega, and in only a few words. Zvelebil's data also include several Kannada-like elements, and it is entirely likely that his consultant may have even spoken a mixed, intermediate form of Solega.

In recent years, the media, researchers, many of whom have affiliations to ATREE and VGKK, and the general public have become increasingly aware of the existence of the Solega, due to their involvement in a land-rights movement and a heated wildlife conservation debate (see Sect. 1.7.10 for a more detailed account). Certain aspects of Solega traditional knowledge have therefore been presented in academic journals and in popular magazines, primarily as a means of showcasing the beneficial impacts of Solega agricultural activities on the forest ecosystem, or the potential for incorporating Solega traditional practices into modern forest management regimes [88]. The Solega's involvement in a participatory resource monitoring programme [89] as well as in organic agriculture and self-help projects [90] have also been reported. Many practical issues being faced by modern Solega have also drawn the attention of researchers. These include the effect on Lantana on the livelihoods of forest-dwelling Solega [91], low crop yields [92] and the interactions between traditional and modern healthcare systems [93].

1.7.3 Governance

Solega society is arranged into five exogamous clans or *kula*, which are arranged in an informal hierarchy. Within each Solega village, one man from each clan is elected to join a council that is responsible for settling disputes and maintaining law and order. Each member of the council has a formal title, and is responsible for a set role (Table 1.1). This is one of the few instances when the clan hierarchy manifests itself in a concrete fashion, as the representative of the *tene:ru* clan, the *paṭṭaga:rã*, is always the head of the council, while the *ejama:na*, elected from the *seḷikiru* clan, is considered his deputy. There appears to be a minor hereditary component in the election of a council member, in that a position seems to be regularly filled by members of a single nuclear family. Nevertheless, a person is elected to council only through the consensus of the inhabitants of a village, and he can also lose his position if public opinion turns against him. Solega society is currently far more egalitarian than it once was. Council members used to command a considerable amount of prestige and authority, and people would move aside to let them pass, while women would avert their gaze out of respect. They were also the sole judicial mechanism that ordinary Solega had recourse to, and had the power to mete out, and enforce, punishments as severe as the complete ostracism of serious offenders from all social contact for long

Table 1.1 Titles of the Solega elders, the clans they are elected from and their official responsibilities

Title	Clan	Responsibility
paṭṭaga:rã	*tene:ru*	The most senior of the elders, head of the council
ejama:na	*seḷikiru*	Deputy to the *paṭṭaga:rã*
maṇega:rã	*ha:laru*	Passing of judgements for punishing crimes
salava:di	*su:rru*	Administrative roles, including the role of sheriff
koluka:rã	*beḷḷuru*	Arresting criminals and meting out physical punishments; also gives names to newborns

periods. Nowadays, council members are still shown the required amount of deference at festivals and other formal occasions, but their temporal power seems to have been somewhat eroded.

1.7.4 Religion

Modern Solega religion is a comfortable mix of animist and Shaivite Hindu traditions, and includes practices ranging from the making of offerings to placate natural forces and entities to the worship of maïnstream Hindu gods in a manner that would be recognisable to the average resident of Bangalore. The holiest site in the religion is an ancient *Michelia champaca* tree, known to the Solega as *doḍḍa sampage*, and situated near the settlements of *Keredimba* and *Gombegallu*, which lie well within a large expanse of evergreen rainforest. While the tree itself is revered as a sacred being, current Solega tradition attributes a great part of the significance of this site to a visit by Shiva and Parvati during the *satyada ka:la* or 'time of truth' back in the ancient past. Accordingly, the base of the tree is adorned with numerous stone *lingam* (the symbol of Shiva), and it is said that a much larger *lingam* is hidden deep below the tree's roots as well. Many Solega also celebrate Kannada festivals such as *yuga:di* 'the new year', *a:yudha pu:je* '*navara:tri* festival, or *dashera* in northern India' and *gaṇe:sh chaturthi* 'Ganesh's birthday'.

The Solega worship five *kula de:varu*, one for each clan, and hold a major festival for each deity every single year. At the start of the agricultural season, Solega farmers make offerings to the elephant god, asking him to spare their lives and their fields. Similarly, a festival called *roṭṭi habba* is held at the time of the *ra:gi* 'millet' harvest, to thank various deities for a good crop. Other forest animals are also revered by the Solega, for their association with various gods—these include *ka:ṭi* (Indian bison or gaur, *Bos gaurus*), *karaḍi* (sloth bear, *Melursus ursinus*) and *huli* (tiger, *Panthera tigris*). In addition, several birds and animals named in the song-cycle *ha:ḍuke*, which is sung from dusk till dawn in festivals, are said to be *de:varu* 'gods'—these include *to:ḷiyakki* 'bats', *soreyakki* 'pigeon/dove', *muḷḷandi* 'porcupine', and so on.

1.7.5 Division of Labour

Domestic tasks, including childrearing, cooking and cleaning are performed almost exclusively by Solega women, whereas most tasks that take place outside the home are shared by the entire family. Examples of the latter include collecting firewood, collecting yams and other edible forest products, preparing land for agriculture and tending to crops. Honey harvesting is carried out by men, either in groups, or singly, while women are usually in charge of grazing livestock, such as goats or cattle, should a family possess any. Solega of both sexes seek employment as labourers or farm-hands in nearby coffee estates, construction sites or towns, or in short-term state government initiatives such as the roadside clearing of the woody weed *Lantana*.

1.7.6 Nutrition

The traditional staple food of the Solega is the finger millet (*Eleusine coracana*) or *ra:gi*, whose flour is cooked with water and moulded into balls or *iṭṭu*, and eaten with a lentil or meat curry (*uduka*). In the past, *ra:gi* was cultivated by practically all Solega families, along with other food crops such as pumpkin, corn, banana and a variety of legumes. The only foodstuffs that were routinely purchased or bartered for in town included salt and cooking oil. Nowadays, many Solega have taken to growing coffee on the plots of land that have been granted to them by the state government. This is seen as an advantageous strategy because the sale of coffee beans provides cash, and the coffee plants themselves are not prime targets for wild animals. In contrast, *ra:gi* fields are regularly raided by wild pigs and birds, which have the potential to reduce harvests by a significant proportion (up to half). Elephants could easily wipe out an entire year's efforts in a single night, and *ra:gi* cultivators routinely build temporary shelters next to their fields to keep watch over their crop.

In recent years, rice has become firmly established as the main staple in many Solega settlements, and parents often complain that children now refuse to eat *ra:gi iṭṭu*. A cash income has allowed many Solega to buy other new kinds of food from nearby towns, instead of having to seek out traditional forest foods. Solega elders blame the increasing consumption of 'city' foods, grown with the help of pesticides and artificial fertilisers, for what they perceive to be a rise in the prevalence of disease in their community. Many older Solega will also emphatically state that the food bought from towns—including *ra:gi* grain for coffee-growing families—has neither the taste nor the aroma of home-grown food.

The Solega were once heavily reliant on forest foods, including various kinds of edible fruit and mushrooms, yams, edible foliage and honey. Many kinds of animals and birds (along with their eggs) were trapped and eaten in the past, but this practice has now stopped, subsequent to the setting up of the wildlife sanctuary. Unsurprisingly, the availability of forest foods is patchy and seasonal, and isolation, a lack of ready cash and external restrictions frequently conspire to make it difficult for Solega families to obtain adequate nutrition. One of my primary consultants recalls that the monsoons in the year of his marriage were particularly active, and that his financial situation was simultaneously dire. Unable to leave the house to find work, or to seek forest foods, he and his young wife had to subsist on their meagre store of yams for a couple of months. This occurred in the late 1990s, and it is entirely likely that many Solega families, especially those living away from roads and other infrastructure, frequently struggle to feed themselves to this day.

1.7.7 Modern Living

Until recently, the Solega tended to actively shun contact with non-Solega, only traveling down to nearby towns to trade honey, grain or lichen for items that the forest did not provide, or to find employment as agricultural labourers. Older Solega still tend to avoid contact with outsiders, however, although the times of Solega people running away and hiding, upon spotting 'plains people' traveling through their forest, have now truly passed. The construction of a state highway, the establishment of a jungle lodge for tourists and the introduction of regular bus services have greatly facilitated the movement of 'plains people' into Solega territory, either as tourists, or as migrants, and this has had a great impact on the lifestyles of the Solega living in settlements close to the road.

The mobile phone revolution that has been sweeping through practically every corner of India has made a definite impact on Solega life. At least three different mobile service providers are currently operational in the B. R. Hills, and it is not uncommon to see Solega elders, even from remote villages, using mobile phones. Many young Solega males, especially those from settlements near the state highway, routinely listen to the local Kannada FM radio station or to downloaded mp3 recordings of Kannada and Hindi songs on their mobile phones, and can also be seen taking photographs or making video recordings with their phone cameras.

Solega families that have made the switch to coffee growing, and have cash incomes, have been able to indulge in urban comforts, such as a power supply, a television set (often accompanied by a DVD player and satellite dish), and the addition of a guest room to their existing dwelling. Many are adopting the customs of their Kannada-speaking neighbours: earlier, marriage was never an occasion that was publically celebrated,[7] but urban, Kannada-style marriages, with music, feasting, a religious ceremony and sometimes even a professional cameraman, are becoming more popular.

[7] A boy and a girl would simply elope, live in the forest for a few months, and then return to their settlement as a married couple.

1.7.8 Socio-economic Situation

The 2001 Census of India reported that out of a total of 3.4 million individuals recognised as belonging to Scheduled Tribes in the state of Karnataka, just under 30,000 individuals were identified as 'Soligaru'.[8] Out of these, around 3300 and 5100 individuals were employed long-term as 'cultivators' and 'agricultural labourers' respectively [94]. A further 2900 individuals worked as agricultural labourers on a 'marginal' basis.[9] Overall literacy levels among the Solega are low, even for the rural part of the state that they primarily inhabit. The literacy rate for Solega above 6 years of age was 33 % in 2001, and was heavily skewed in favour of males (M: 39 %; F: 26 %). In contrast, the literacy level for the state of Karnataka was 67 % (M: 76 %; F: 57 %) [95], while the rates for the Yelandur and Chamarajanagar subdistricts, the populations of which are >80 % rural, were 50 % and 49 % respectively [96]. The overall literacy rate for all Karnataka Scheduled Tribes combined was 48 % (M: 60 %; F: 37 %) [94, 97], markedly better than that of the Solega.

1.7.9 Attitudes towards Language and Traditional Knowledge

The Solega language has little prestige outside the community, and the Solega's own conceptions of the relationship between their own language and Kannada are varied. I have heard the language referred to as *soliganuḍi* 'Soliga speech' by Kannada speakers, who tend to also classify Solega as a dialect of Kannada. The Solega sometimes echo this categorisation by sometimes referring to their speech as *namma kannaḍa* 'our Kannada', although, for the most part, the generic phrase *namma ba:se* 'our language' is used. Solega speakers are aware of the proximity of their language to Kannada, and can, in most cases, alter their speech so as to make it intelligible to Kannada speakers. At the same time, they point out that Kannada speakers would find it impossible to understand a normal-speed conversation, in 'pure' Solega, between speakers from interior settlements.[10]

Practically all Solega recognise that their language is changing, and that 'young people' no longer speak like their parents or grandparents. Language attrition in favour of Kannada was prevalent in all settlements, but was strongest in lowland villages, and weakest in interior villages. Solega living in certain lowland settle-

---

[8] *Soligaru* is the Kannada plural form for *soliga*. The 2001 census recorded a further 124 individuals, who were identified as 'Sholaga', and provided separate data for them. In the absence of any further explication of the difference between the two groups, I would suggest that they are the same.

[9] 'Marginal workers' is the official English gloss for the Hindi term अल्पकालिक कर्मी *alpaka:lik karmi:*, and can be understood to mean people who are employed on a casual or part-time basis.

[10] I have played recordings of Solega, as spoken in interior settlements, to Kannada speakers from Bangalore. These people found some of the vocabulary and grammar to be familiar, but struggled to comprehend the content of the recordings to any great extent.

ments have now completely switched to Kannada, while in the highland villages, children of school age (and young adults) would be able to speak little or no Solega, but still have a passive comprehension of it. Their parents, now in their 30s and 40s are still able to hold an extended conversation in 'pure' Solega, but would, in practice, frequently switch between Kannada and Solega. Even in the highland villages near the field station (i.e. accessible by road, but within the rainforest proper), certain high frequency Kannada words, such as *benki* 'fire', have completely displaced their indigenous Solega counterparts—*kiccu* in this case. People were definitely aware of the Solega word, but only used it in my presence for humorous effect. It is only in the interior settlements that very young children can still speak Solega as a first language. The division of labour between Solega and Kannada among bilinguals seems, on the surface, to be straightforward. Simply put, Kannada is used with non-Solega people, while Solega is reserved for in-group members. Most adult Solega in their 30s and 40s are now bilingual, however, and are equally at home in either code. I did observe Solega bilinguals frequently switching between codes, but neglected to record any sociolinguistic contexts that might motivate the switches.

The knowledge of ritual texts, plant and animal names, their uses and usual habitats, place names and their cultural significance, and other items of specialised vocabulary and cultural information is being significantly eroded in all but the interior villages. My consultants, with whom I had discussions on this topic, regretted this state of affairs, but were pragmatic about their children's opportunities to keep up the old ways. These consultants, in their 30s and 40s, arguably represent the last Solega generation to have been brought up in the times when shifting cultivation was still practised, but was already on the decline. Now faced with a very different reality, this generation of Solega are convinced that the only way for their children to prosper (or at the very least, escape crushing poverty) would be to be educated in, and adopt, mainstream ways of living. Another extremely significant factor that has precipitated a break in the transmission of traditional knowledge is the dominance of *Lantana camara* in the forests of the B. R. Hills. Acting as a cruel and highly visible metaphor for the influx of foreign cultural influences into the Solegas' world, this plant has completely swamped large tracts of the forest understorey, obliterating old forest trails, swallowing up sites of religious and utilitarian significance, and driving many culturally important species of plants and animals to local extinction. The impact of this invasion on Solega cultural transmission is as noxious as any government directive to curtail traditional activities—*Lantana* ensures that the knowledge of countless plants and animals will fade from the Solega's collective memory, even as those species quietly disappear from the forest floor.

1.7.10 Attitudes Towards the Forest

"*The forest is our home*", is a statement that all Solega will, no doubt, agree on. Far from being a trivial statement of fact, it is an utterance that also conveys a strong sense of pathos, for the Solega are no longer free to live in their home as their ancestors did.

Government restrictions prevent the Solega from hunting, collecting timber and setting fires to clear land for agriculture, while *Lantana* has made it difficult for people to travel, as well as find important medicinal plants and traditional foods. Perhaps the most alarming impact is the change in the behaviour of elephants, who now find it hard to locate the grasses (particularly *ba:ṇe hullu*) that dominate their diet, and are subsequently more ill-tempered, unpredictable and violent. Dense *Lantana* thickets also obscure the vision of people travelling through the forest, and prevent the early detection of dangerous wild animals, including elephants, bears and wild pigs.

Given that the forest ecosystem is visibly in decline, the Solega are angry that their own traditional activities have been curtailed, as they see a very clear link between their agricultural practices and the health of the forest (see Chap. 5 for a more detailed discussion). Briefly, the clearing of the forest understory and accumulated leaf litter through controlled burning has the double advantage of keeping *Lantana* in check and promoting the growth of the seedlings of important rainforest trees. Many older Solega remember times when *Lantana* was scarce and wild animals and birds were plentiful, as was a diverse assortment of understorey shrubs and herbs, edible mushrooms and honey with the aroma of now-rare wildflowers. The exact nature of the link between agricultural fires and *Lantana* prevalence remains to be determined empirically, and is an important issue that is still being investigated [98].

Two momentous policy decisions were made by at the national and local government level in recent years, that have the capacity to radically alter the future trajectory of Solega habitation in their ancestral forests. The first was the Recognition of Forest Rights Act (RFRA), 2006, which went some way towards reversing the highly proctectionist tone of its predecessors, the Indian Forest Act, 1927 and the Wildlife Protection Act, 1972. The RFRA granted 'forest dwelling Scheduled Tribes and other traditional forest dwellers' the right to live on, and cultivate, their traditional land, collect minor forest produce, and also be granted legal title to land that was being cultivated prior to a cut-off date. As can be expected, this development was warmly welcomed by the Solega, as many became the undisputed owners of small parcels of land, and small-scale commercial activities, such as the collection of honey and lichen for sale through the *Soliga Abhivruddhi Sangha* (SAS) 'Soliga Development Co-operative' could recommence.

The second, and more recent, policy directive to reach the B. R. Hills is the decision of the Karnataka state Forest Department to notify that region as a Project Tiger Reserve. Tiger Reserves are traditionally people-free zones, and in the past, the inhabitants of areas so designated were either forcibly evicted, or convinced to leave by means of monetary enticements. Prominent Solega elders and activists, supported by organisations such as the SAS, ATREE and VGKK have publically and unanimously opposed the creation of a Reserve in the traditional sense, and have instead proposed a more inclusive 'community-based tiger conservation model' [99].

Chapter 2
Ethnotaxonomies and Universals: Investigating some Key Assumptions

2.1 Introduction

It is often stated that the study of ethnotaxonomies have great potential to inform theories on human language in general, and, by extension, human cognition as a whole. It is claimed that there are strong regularities among the folk taxonomies of the world's languages, and that this is due to similarities in the way humans perceive the biological world in which they are perpetually immersed. Scott Atran [100] is one of the strongest proponents of the theory that *"universal taxonomic structures universally constrain and guide inferences about the biological world"*. Drawing on evidence from the folk taxonomies of unrelated languages, and from psychological experiments, Atran concludes that:

> The uniform structure of taxonomic knowledge, under diverse sociocultural learning conditions, arguably results from domain-specific cognitive processes that are panhuman, although circumstances trigger and condition the stable structure acquired...
>
> Although accounts of actual causal mechanisms and relations among taxa vary across cultures, abstract taxonomic structure is universal and actual taxonomies are often recognizably ancient and stable. This suggests that such taxonomies are products of an autonomous, natural classification scheme of the human mind... Such taxonomies plausibly represent "modular" habits of the mind, naturally selected to capture recurrent habits of the world relevant to hominid survival in ancestral environments.

Such sentiments have repeatedly been voiced in the past in the context of many different aspects of human language and cognition—the most prominent among these is, of course, Noam Chomsky's Universal Grammar, which stresses the innateness and universality of a set of grammatical rules, which are meant to reside in a discrete module of the human mind. The notion of universals has met with considerable objection from field linguists, who routinely find, among their lesser-known languages, counterexamples to the universalists' homogenising claims (for a recent review, see Evans and Levinson [101]). Even linguists who take the existence of universals for granted now distinguish between 'unrestricted' universals (of the form "All languages have X") and 'implicational' universals (which are phrased as

A. Si, *The Traditional Ecological Knowledge of the Solega*, Ethnobiology,
DOI 10.1007/978-3-319-24681-9_2

"If a language has X, then it also has Y") [102]. Although the existence of implicational universals has been known to linguists for several decades (see [103] for a first elucidation of the concept), systematic comparison of hundreds of languages from around the world has necessitated *"a rethinking of the nature of Universal Grammar, which is usually thought of as a set of unrestricted universals"* ([102], p. 13). Croft explains this point further:

> The part of Universal Grammar that consists of unrestricted universals specifying ways in which all languages are identical captures only a very small portion of what is universal about language. It misses most of what is universal about language beyond the basic design features.[1] The presence of large numbers of implicational universals requires a model which allows for a great deal of variation in grammatical structure across languages, but constrains that variation to a significant degree in many different dimensions... The second general conclusion about language universals that is revealed by typological research is that the constraints on language variation that are represented by implicational universals are not exceptionless[2] in the way that the design features of language are. (p. 13–14)

Universal patterns have also been proposed for the semantic domains of colour, landscape and body-part terminology, but only a few have survived careful scrutiny in any way [104, 105]. In their first major publication on colour terminology in the world's languages, Berlin and Kay [106] made strong claims regarding the existence of universal patterns in the cross-linguistic naming of colour terms. Some later authors, such as Dedrick [107], have rephrased these universals as 'strong tendencies' (p. 106), while Lucy ([108], p. 323) has objected to the near-exclusive use of Munsell arrays in the colour naming tasks. This, he points out, ignores critically relevant information, such as the normal everyday referential uses of the colour terms and the grammatical status of these terms in a language. Using very different methodology, Wierzbicka [109] argued that the (English) concept of 'colour' has no place in the semantics of the Australian language Burarra; the two 'colour' words described in this language refer to more global visual categories, for which brilliance is at least as importance as hue. The issue of colour universals continues to be debated (e.g. [110]), although a middle-ground consensus seems to be emerging [111, 112], which recognises both universal constraints (albeit weaker), as well as an effect of language on colour categories.

2.2 Universals in Folk Biological Classification

Given this context, it is surprising that the existence of universals in another well-studied domain—the folk classification of plants and animals—is, to an extent, taken for granted by most ethnobiologists, with the exception of a dissenting

[1] These are very basic features that include: utterances made up of discrete meaningful units; conventions of syntax; and duality of patterning (independence of sound structure from syntactic structure) [102].

[2] An example of an implicational universal presented in Croft [102] is the following: If a language has Adjective–Noun word order, then it (almost always) has Numeral–Noun word order. Note that a violation of this 'universal' is possible, and even attested, in a small number of languages.

minority. Although many models of ethnobiological classification have been proposed in the past, the most influential by far has been the hypothesis presented in the volume *Ethnobiological Classification* by Brent Berlin [9], which is a reformulation of an earlier proposal by Berlin, Breedlove and Raven [12]. The 1992 work has directly influenced much of the contemporary research effort on folk biological classification, and it is worth considering some of its main predictions—these are briefly described below.

2.2.1 Ethnobiological Classification by Berlin

Berlin's model seeks to describe not only the manner in which people categorise the living organisms around them, but also the constraints on the nomenclature of the resulting groupings. Following are the main features of the model:

Categorisation

Only a subset of the plants and animals within a locality are named by the humans who live in that locality.

1. Categorisation is based primarily based on observed morphological and behavioural affinities and differences.
2. Taxa are grouped inclusively into a hierarchic (taxonomic) structure.
3. Taxa are spread out over six ranks or levels, whose content is comparable to the ranks of western biological classification. The six folk ranks are kingdom, life form, intermediate, generic, specific and varietal.
4. Across all folk systems of ethnobiological classification, taxa of each rank exhibit systematic similarities in their relative numbers and biological content.
5. Some members of generic and subgeneric taxa are thought of as being prototypical (the same may be true of life form and intermediate taxa).
6. A substantial majority of folk taxa corresponds closely in content to scientific taxa. This is especially true for taxa of generic rank.

Nomenclature

7. Intermediate and kingdom-level taxa are often not named (covert) in folk systems. Some life form taxa may also be covert.
8. There are two kinds of names—primary and secondary. Primary names can be simple (*dog*) or complex (*forget-me-not*), while secondary names are complex (*blue gum*), and contain the name of an immediately higher taxon (*gum*). Complex primary names can be productive (*catfish*), or unproductive (*silverfish*). Secondary names form part of a contrast set (*blue gum* and *river gum*).
9. Generic, life form and intermediate taxa are labelled by primary names, while subgeneric taxa (with notable exceptions) are labelled by secondary names.

10. A subgeneric taxon is given a primary name if it is either the prototype of the genus, or of major cultural importance.
11. Names commonly allude to characteristic traits of their referents.

Other Issues

12. There is an 'evolutionary sequence' for the appearance of named taxa: generic taxa are named first, followed by higher order taxa, and finally subgeneric and kingdom level taxa. The recognition of subgenerics is directly tied to plant domestication.

Even a cursory scan of the above predictions reveals strong parallels with the research efforts that sought to uncover linguistic universals in colour terminology. The concept of prototypical categories, the emphasis on visual salience alone as the driving force behind categorisation, and the 'evolution' of the domain-specific lexicon from simple to complex as a mirror of a culture's level of technological development are all ideas that have been discussed previously. Add to this the methodological framework used in Berlin [9]—the near exclusive use of prepared specimens as visual stimuli to be identified, named and categorised by linguistic consultants—and it becomes clear that the study of folk biological classifications has been carried out with much the same assumptions that guided research into colour classification and body part terminology.

Two major lines of questioning can be followed at this point. (1) Can the general principles listed above really be considered ethnobiological universals? Many, but not all, of these principles are examined more fully in the following two chapters, in the context of Solega naming and classificatory practices relating to plants and birds. It will be shown that the Solega data repeatedly fail to support many proposed 'universals' of ethnobiological classification and nomenclature. (2) Is it reasonable to expect the existence of 'universals' based on the arguments made in Berlin [9]? This question is addressed in the current chapter, which investigates the theoretical foundations for the claim that all human languages (along with scientific taxonomy) classify living things in much the same way. In the following sections, I will argue that many of these basic assumptions, as they relate to the Berlin model, are flawed, as a result of which incorrect conclusions have been reached about the nature of ethnobiological classification.

2.3 On Challenging and Defending 'Universals'

'Universals' are easy to posit, but hard to defend. Many absolute linguistic universals have been proposed, for which counterexamples can be found—often from small, recently-described languages [101]. Evans and Levinson explain that less than 10 % of the world's living languages have adequate descriptions (a full grammar and dictionary), and that making generalizations over such a small sample size is a very risky business:

> The fact is that at this stage of linguistic inquiry, almost every new language that comes under the microscope reveals unanticipated new features (p. 432).

The number of languages for which adequate ethnobiological studies have been published is likely to be vanishingly small in comparison (e.g. the survey of 39 languages in [113]), and the task of formulating meaningful generalizations correspondingly more difficult. The fact that unrelated languages from many different parts of the world behave in a similar fashion is interesting, and perhaps indicative of a deeper phenomenon worthy of investigation, but the smaller the initial sample size, the more likely that the observed similarity will turn out to be a trend, and not a 'universal'.

The task of challenging an absolute 'universal' is an easy one. Strictly speaking, all it would take is a single, verifiable counterexample to bring the entire edifice of the 'universal' come crashing down. For after all, what are 'universals', if not testable hypotheses? The person seeking to prolong the life of a 'universal', on the other hand, has the unenviable task of having to defend the idea in question against a potentially unforeseen piece of empirical data. Two options are available to such a person: address the challenge directly, and argue why it is not a valid refutation of the 'universal', or modify the 'universal' (i.e. the theory) in accordance with the new data. It is *not* a valid strategy to point out the existence of published studies that support the 'universal' as defence against the new data. Those studies are the foundation that the 'universal' was built upon, and the introduction of new data that do not conform to the predictions made by the 'universal' becomes an issue that has to be dealt with in its own right.

2.4 Problems with Berlin's *Ethnobiological Classification*

Many of the general principles proposed by Berlin in his 1992 monograph have resonated with ethnobiologists working in different parts of the world, as is evidenced by the many scholarly publications purporting to lend support to certain 'universals'. The following discussion (along with further tests of Berlin's predictions later in the book) is *not* meant to suggest that Berlin's general principles will never apply to any language. On the contrary, I believe that the general principles likely capture some important truths about how languages will often converge on certain solutions in spite of being unrelated. What *is* being argued here is that these general principles are by no means universal (i.e. there will be languages that violate one or more of the general principles), and this is made abundantly clear by the data presented in Chaps. 3 and 4. A more immediate concern of this section is the theoretical justification underlying the existence of general principles in the first place. There may well be identifiable reasons that explain why many languages behave in similar ways with regard to ethnoclassification and nomenclature, but I do not believe that they are identified by Berlin.

One of the key proposals of the Berlin model is that ethnobiological classifications worldwide share broad, predictable similarities, and that such classifications are, in turn, surprisingly close to those produced by professional scientific taxonomists. Such a concordance, Berlin claims, cannot be coincidental, and must therefore be the result of shared, innate properties of human cognition—scientists and non-scientists alike perceive the living world in a very particular way, and this results in very similar classification schemes. As preliminary evidence for this claim, Berlin discusses the results of the impromptu demonstration experiments he regularly holds in his undergraduate classes. A student with no biological training is called to the front of the class, and shown a pile of museum specimens of birds from Berlin's field site. He is then asked to arrange them in whatever way he deems appropriate.

> The student's efforts always result in a series of neatly stacked groups of individual birds, usually lined up in a row. The piles correspond perfectly to the groupings recognised by scientific ornithologists, as well as to those of the Huambisa and Aguaruna Jivaro from whom the specimens were collected … This informal exercise … points out several facts about how humans discern the natural clusters and clumps of biological reality—clusters that are perceivable, one might say, from *distinct perspectives*, or with different *degrees of resolution*. (p. 10)

> To the extent that the Western scientist and the native ethnozoologist see the *same organisms*, and to the extent that, without prior discussion of the local system of classification, the former's classification of the local fauna turns out to correspond closely to the latter's as regards which species are singled out for naming, then one may reasonably infer that in both systems grouping is based on the recognition of relative degrees of similarity and difference among species. Since the two naturalists are classifying exactly the same animals, the taxonomy of the Western scientist should be nearly identical to that of our indigenous ethnozoologist. (p. 82)

While the results described by Berlin are, at first glance, compelling, they are based on two critical assumptions which are at least highly contentious, if not deeply flawed. The first is that there is a single "natural order" of plants and animals that it is possible to discover through objective, scientific means. The second is that the work of (mainly European) taxonomists over the last two centuries has indeed been objective, and has resulted in truly natural groupings that adequately capture "*the natural clusters and clumps of biological reality*".

> Since, in any local habitat, biological reality is *not* a continuum but a series of readily definable chunks that can be described in terms of the objective methods of biological field botany and zoology, one is motivated to discover what portions of this reality are cognitively recognised in any particular folk biological system and why. (p. 13)

Taken together, these assumptions give the distinct impression that the practice of taxonomy (in the western scientific sense) involves the simple collection of objective natural facts, and the subsequent arrangement of these facts to produce a single classification scheme that not only reflects biological truisms (and therefore does not change with time—at least within a human lifespan), but also has the consensus of all taxonomists. In the following sections, I will briefly discuss the objections that many researchers have raised about these assumptions.

2.4.1 Many Possible Classification Schemes

While it is unlikely that anyone would ever place oaks and robins in the same taxonomic category, the process of scientific classification is not always as straightforward, and existing groupings are constantly challenged as new data and/or analytical techniques become available. The "biological species concept" of Ernst Mayr, which is frequently cited by Berlin, is currently the most influential of all species concepts, but it is by no means the only one. Ereshefsky [114] describes no less than seven prominent species concepts,[3] which can be placed into three broad categories on the basis of the criteria they use to determine what constitutes a species: interbreeding, ecology and phylogeny. A description of each of these concepts would be beyond the scope of this chapter, but it is important to note that the use of different species concepts can radically change the nature of the classification system: cottonwoods and balsam poplars, for instance, can interbreed, despite their being recognised as different species on the basis of their genetic, morphological and ecological distinctiveness. For Ereshefsky, all three types of criteria stem from legitimate biological phenomena, which is why he believes that a plurality of valid classification systems is possible:

> Because biologists disagree on the correct approach to species, they provide different classifications of the organic world. Moreover, these approaches cross-classify the world's organisms by placing the same organisms in different species taxa. (p. 133)

In addition, he takes his argument even further, and provides reasons why he is not only a "taxonomic pluralist", but also a "metaphysical pluralist", i.e. someone who believes "*that the world is carved in multiple ways, and the most fruitful way to represent that world is with a plurality of classifications*" (p. 43). Species pluralism, then, is a real feature of the world, and is the result of the interplay of the three evolutionary forces named above. While not everyone shares Ereshefsky's opinion, other researchers may still be characterised as taxonomic pluralists, who are simultaneously "metaphysical monists" or "metaphysical agnostics", to use his terminology. Agapow et al. [116], for instance, advocate an intuitive idea of a species as "*a population on an autonomous historical trajectory*", but nevertheless admit that

> ...the very term "species" is deeply ambiguous. While biologists customarily treat species as tangible entities—at least more so than any other sub- or supraspecific group... there is nonetheless a vast spectrum of meanings attached to the word. The argument over how species should be defined is endless... [the] concepts encompass many operational and empirical definitions, often resulting in a given group of organisms being viewed in drastically different ways by different workers.

The phylogenetic species concept, in particular, which currently has equally vocal and passionate supporters and detractors, has the potential to completely alter the way in which species are perceived by taxonomists. In a meta-analysis of almost 90 phylogenetic studies on taxa ranging from lichens to whales, Agapow et al. [116]

[3] Mayden [115] lists over 20 species concepts.

found that the application of the phylogenetic species concept (in place of more traditional concepts based on interbreeding and morphology) increased the species count by nearly 50 % on average. A breakdown of the results by taxon is enlightening; the important point here is that it is not only the groups frequently left out of folk taxonomies that are affected (fungi, lichens), but also those that play very central roles in ethnobiological classifications (plants, reptiles, mammals, birds).

> It is difficult to detect any trend across groups, and given the small sample size and how it was obtained, any putative trends should be treated with suspicion. With this in mind, the greatest increase in species numbers was seen in fungus, a 300% increase… lichens (259% increase) and plants (146% increase). Among animals, reptiles showed the greatest increase (137%). It is startling that taxonomically well-studied groups like mammals, arthropods, and birds showed large and roughly similar increases (87%, 77%, and 88%, respectively).

Regardless of whether or not one agrees with Ereshefsky's viewpoint, the actual practice of taxonomy very rarely results in the production of classificatory schemes that go unchallenged for long periods of time. A brief look at the contents page of any scientific journal dealing with taxonomic issues will confirm this assertion—as new data (the discovery of new species in little-explored parts of the world) arise from fieldwork, and new analytic (genetic and molecular) techniques are developed in laboratories, amendments are frequently made to all levels of the taxonomic hierarchy, from Kingdom (e.g. [117]) to Species (e.g. [118, 119]). Note that the latter two examples pertain not to obscure, hard-to-perceive taxa, but to two groups that Berlin would definitely call perceptually salient—fur seals/sea lions and edible yams. Even the biological species concept has come under attack in recent years, both for conceptual reasons (which will not be discussed here), as well as practical ones:

> Although the biological species concept provides an unambiguous criterion for differentiating and delimiting species, one of the major problems is the practical impossibility of ascertaining reproductive isolation between large numbers of populations in the wild. As a result, most taxonomists, even those that accept the biological species concept, continue to use morphology and other phenotypic characters in order to delineate species boundaries ([120], p. 690).

Balakrishnan goes on to note that the competition between rival species concepts is unlikely to be resolved in the near future, particularly because biological systematics aims at achieving two tasks that are often in opposition: that of reconstructing the historical paths of evolution, and that of providing a practical frame of reference for other biological studies. One may therefore object on philosophical grounds alone to the desirability of having a universal species concept imposed on all the subfields of biological research.

2.4.2 *'Objective' Scientific Taxonomy*

An issue related to the one discussed above is that of the accuracy and objectivity of "modern" scientific classifications, something Berlin seems to take for granted. In saying that the folk classification of a particular language closely approximates scientific taxonomy, and that this observation indicates a universal human propensity to recognise real divisions in nature, Berlin assumes that there is something absolute about the relationships between the binomial Latin names that have dominated western science for the last two centuries. Berlin also points to a phenomenon called the "hollow curve" in scientific taxonomy, where most plant and animal higher taxa contain large numbers of very small (often monotypic) genera, and a few very large genera. This, he claims, is identical to the pattern found in folk taxonomies, and lends support to his hypothesis that all humans perceive and classify natural divisions in similar ways. This view has been challenged in recent years, not only by supporters of the phylogenetic species concept mentioned above, but also by more traditional taxonomists such as S. M. Walters, former Director of the Cambridge University Botanic Garden. Walters [121] is remarkably forthright about his attitude towards traditional Linnaean classification, saying that the classification of flowering plants, at least, suffers from a "European bias" that is largely ignored by other taxonomists.

Drawing from a number of historical sources, and from the writings of contemporary taxonomists, Walters argues that the hollow curve is actually an artefact of the attachment of conservative taxonomists to the categories first proposed by Linnaeus. These categories were, after all, heavily influenced or even based directly on pre-existing European folk taxa. All of the 20 largest flowering plant families in existence today were first recognised by Linnaeus or by his predecessor Tournefort, and were based on plants that were either common or culturally important in Europe—these include plants such as the rose, daisy, nightshade, carrot, lily, mint, and so on. Walters' argument is that as specimens of new species poured into Europe from around the world, taxonomists loyally tried to group these species into familiar Linnaean (and thus, European) genera, leading to an artificial inflation of species number in these old genera.

> Any group of flowering plants possessing some obvious 'natural resemblance', which contains a number of common, widespread European species of importance in the life of the community, is likely to contain an 'excess' of genera and *ipso facto* constitutes a relatively large modern family. In addition to the carrot family, Umbelliferae, familiar examples are the mint family, the Labiatae, and the cabbage family, Cruciferae. It has long been familiar to both professional taxonomists and the users of Floras written by those professionals that these three families are peculiar in that it is usually possible to say that any given plant does or does not belong to one of these families but it is often very difficult to decide in which genus within the particular family it should be placed. Putting it crudely, the three families are 'good' or 'natural', the genera within them are 'bad' or 'artificial'. Such families were established in classical and Medieval European folk taxonomy, and the names of their constituents in the classical languages became the modern genera (p. 6).

Recent examples of widespread revision of generic boundaries in these venerable families are not hard to find, especially outside the Old World. Mitchell [122], for instance, presents a heavily revised taxonomy of New Zealand Brassicaceae (the current name for Cruciferae), noting that the taxonomic history of this group of plants reflects the influence of Northern Hemisphere taxonomists. In particular, he mentions that "*most of the New Zealand species were initially placed in well-known or common Northern Hemisphere genera, but as better quality and more plant material became available, these species were placed in new and often exclusively New Zealand genera*". Such changes did not take place until the late 1980s.

Utilitarian biases can also be found in the roots of scientific taxonomy. Both grasses (Graminae) and sedges (Cyperaceae) are widespread and ecologically important in Europe, and both groups are difficult to identify, because of their small flowers [123]. However, grasses are extremely important as food crops or as fodder, whereas the sedges are not. The result was that Linnaeus created 30 small genera of grasses (based on folk taxa), compared to a single sedge genus *Carex*, in which were included 29 species. Ironically, the historical explanation also accounts for the profusion of monotypic genera seen across all plant and animal groups—as new species were discovered, they were placed in the old Linnaean families, so as to cause minimal disruption to that venerable classification scheme. Both Walters [123] and Clayton [124] provide good historical evidence that the creation (by taxonomists) of most of the flowering plant monotypic genera occurred in post-Linnaean times. Also drawing on Willis' [125] "hollow curve" data, Walters provides an illuminating example: of the genera in one subgroup of the buttercup family (Ranunculaceae), six of the eight most species-rich genera (having 10 to 300 species each) are found in Europe, whereas 23 of the remaining 25 genera with less than 10 species each (14 being monotypic genera) are from Asia, Africa and the New World.

Walters is not alone in claiming that certain aspects of scientific classification have little to do with real divisions in the natural world:

> It emerges that so far no biological counterpart has been found for the genus… Its formal definition remains substantially in its original form as an abstract and arbitrary level of morphological difference.
>
> Nevertheless, though its biological meaning is far from clear, the genus is deeply entrenched in taxonomic thinking, and few biologists would attempt to deny its existence. Nor is it purely a construct of modern science, for many of our present genera have been taken almost unchanged from traditional folk taxonomies ([124], p. 150).

Most importantly, Clayton asserts that many taxonomists continue to add taxa to pre-existing classifications using decidedly non-biological criteria. Taxonomic revisions based on new evidence are also slow to take hold—the information in local floras and field guides therefore needs to be used with a great deal of caution.

> …There is an obvious conflict between the theoretical acceptance of large genera, and mnemonic rejection of them… the outcome tends to be weighted in favour of convenience, appreciating the practical advantage of simple circumscription and easy identification.

> …The conclusion that convenience, in the sense of possessing useful mnemonic and discriminatory properties, is a necessary ingredient of the genus concept may be unwelcome to the purist. However it is worth reflecting upon the divergent treatments that widespread genera commonly receive in local floras, whose authors are unaware of the full range of variation, or the reappraisal of generic limits that must often follow the discovery of new species. Such examples demonstrate that our perception of a genus is disconcertingly sensitive to the sample of species available, a circumstance which renders a quest for total objectivity somewhat unrealistic. (p.151)

Cronk [126, 127] takes a more conciliatory view regarding the nature of plant genera, and of the disproportionately large numbers of monotypic genera in modern classifications. While concluding that both 'real' and 'artificial' components lie hidden in modern classifications in equal proportions, he nevertheless concedes that grouping species into oversized genera and monotypic genera provides little information regarding the species' evolutionary connections, thereby making plant taxonomy "irredeemably inefficient". Stating the case of the large and economically important plant family, the Leguminosae, Cronk [126], like Walters before him, demonstrates that the present-day taxonomy of the three sub-families, Caesalpinoidae, Papilionoidae and Mimosoidae closely follows the patterns established by Linnaeus and continued by the great botanist George Bentham in 1865. Accordingly, Caesalpinoidae contains the smallest genera and Mimosoidae the largest, reflecting the fact that many species of the former group were known to Linnaeus. In contrast, Linnaeus had access to very few specimens of Mimosoids, leading him to group them together under one genus.

In his discussion on the development of the concept of "genus" in scientific taxonomy, Berlin attempts to draw links between folk generic taxa and scientific genera by citing a "*perceptive paper*" by Bartlett [128]. Here, Berlin says, Bartlett outlines the parallels between scientific and folk genera. However, using data from historical sources, Bartlett makes the following observation about the internal structure of scientific genera:

> ...the inclusiveness or size of genera, now as in the past, is less a matter of science than of linguistic preference and convenience.

> ... Greene has shown that the generic idea "oak", as held today, was really borrowed by scientific systematic botany from the folk science of the English pioneer settlers in temperate America, who extended the English folk concept of "oak" to cover the various widely different American oaks. In the eastern United States we distinguish white oak, burr oak, chestnut oak, live oak, scarlet oak, black oak, shingle oak and others, having a perfectly binomial nomenclature for them in English, and, from the literary record, we may be sure that these designations owe nothing to scientific botany. (p. 351)

One must wonder how much "borrowing" from folk systems went on during the time when field naturalists and collectors were scouring "exotic" locations around the world for specimens, and were heavily reliant on the input and knowledge of their native guides. Whether this could account for the perceived similarities between folk and scientific taxonomies would be hard to verify, but it is nevertheless a possibility that cannot be ignored.

The practice of taxonomy in western science has always been strongly influenced by prevailing evolutionary theory (or lack thereof) and perceptions of the natural order of living organisms. The Linnaean system was conceived of in the tradition of Aristotelian essentialism, "*not because he observed essences in nature... but because he believed that is how God created nature*" [114]. As noted above, existing European taxa and morphological features were used wholesale in the naming of Linnaean taxa. Two centuries later, Ernst Mayr championed the biological species concept [129], characterised by the presence of interbreeding. Although he advocated that genera be kept of equal size as far as possible, the "hollow curve" has become a dominant feature of modern taxonomy. The advent of molecular phylogenetic techniques in the last two decades now threatens to bring about a revolution in scientific taxonomy, with proponents of phylogenetic classification demanding that the old Linnaean scheme be replaced with a new PhyloCode [130]. Unfortunately, by constantly referring to "Western science" or "scientific taxonomy" as though these endeavours were totally unbiased and free of disagreement, Berlin paints an oversimplified view of the scientific process that tends to be misleading. Such a view glosses over inconvenient facts that show scientific taxonomy and systematics to be a highly contentious field. Researchers who carry out meta-analyses of taxonomic papers have repeatedly pointed out that the geographic distribution of taxonomists and their research interests are highly skewed, as a result of which countless important species and habitats remain understudied [131], and by consequence, inappropriately classified. This is particularly true of tropical species.

2.5 Synthesis

In the preceding sections, I have tried to argue that the goal of constructing a single scientific classification scheme for a single natural order is at the very least contentious, if not totally unfeasible. Taxonomists are deeply divided on the best way to develop even a classification for practical purposes, let alone one that faithfully represents real divisions in the natural world. For much of the last 200 years, the business of taxonomy has been dominated by an exclusive focus on the morphological features of plants and animals. It is only in the last two decades or so that genetic and molecular techniques have started to be used, either in isolation, or in conjunction with morphological data. This has forced taxonomists to revise large numbers of previously created groupings. Such revisions are slow to take effect, and many of the old names are still used in field guides and floras.

It has been argued that the Linnaean classification system suffers from serious ethnocentric biases, and that utilitarian concerns in European folk taxonomy have motivated some of the more important groupings. Moreover, the kinds of tasks used to elicit folk taxonomies suffer from a range of methodological issues, which make it almost inevitable that the native speakers being interviewed will only use morphological criteria—which happen to be the criteria used in traditional taxonomy. No wonder, then, that folk taxonomists are sometimes found to produce the same categories as scientific taxonomists, leading to superficial similarities between different systems. This contention is amply supported by a key experimental study by Boster and Johnson [56]—two groups of people, "novices" and "expert fishermen" were asked to categorise pictures of fish outlines into as many groups as they wished, and using whatever criteria they wished. The researchers found that the novices produced groupings solely on the basis of morphological criteria, which happened to closely match scientific taxonomy. The expert fishermen, on the other hand, not only showed more variation in their groupings, but also used a combination of morphological and functional (such as good sport, high meat value) criteria. Berlin's "student experiment" can be explained in these terms, and the only generalisation that can be made is that *when humans categorise living organisms using morphological criteria alone, they do so in similar ways.*

The above finding is not particularly remarkable, as the ability to categorise novel stimuli by visual features has been demonstrated repeatedly in human infants [132], many non-human vertebrates [133, 134] and even insects [135]—these include abstract features like number, symmetry, shape, orientation, "sameness" and "analogy". Recently, it was shown that honeybees can even be trained to distinguish face-like stimuli from non-face-like stimuli [136], but no one would ever argue that these were categories relevant to a honeybee. Interestingly, the infant categorisation literature demonstrates that children as young as 20 months can "*use a common name to categorize objects even when those objects were completely dissimilar at the perceptual level*" (reviewed in [137]), and that verbal labels provided by the experimenter can disrupt the formation of perceptually salient categories, when they correlate poorly with the visual category information presented to the infants [138]. Left to their own devices, then, infants will use their innate perceptual abilities to categorise novel objects, but this tendency can be easily overruled by linguistic input. Results such as these and the Boster and Johnson fish experiment show that cognitive processes and a morphology-based folk taxonomy at best convey only half the story, and that the other half must be something other than morphology—something that is often loosely glossed as "utilitarian concerns" or "culture".[4]

2.6 Folk Genera, Rank and Nomenclature

In his discussion on ways to predict and identify generic taxa in folk taxonomies, Berlin makes frequent reference to the scientific concept of genera. He appears to favour Cain's [139] characterisation of the genus early in the history of scientific taxonomy as "*the smallest 'kind' of plant or animal that can be recognized* without close study". Building on this idea, and the fact that many Linnaean genera arose from European folk taxa, Berlin concludes that

> the genus is seen as a configurational category, recognizable almost instantaneously, in contrast with that of subgeneric groupings whose differentiation often requires deliberate and conscious effort to distinguish. (p. 61)

A similar point is made for generic taxa in folk classifications:

> …in the categorization of plants and animals by people living in traditional societies, there exists a specifiable and partially predictable set of plant and animal taxa that represent the smallest fundamental biological discontinuities easily recognized in any particular habitat. This large but finite set of taxa is special in each system in that its members stand out as beacons on the landscape of biological reality, figuratively crying out to be named. These groupings are the generic taxa of all such systems of ethnobiological classification, and their names are precisely the names of common speech. (p. 53)

[4] A discussion on what is meant by "culture" is not possible here, but I would suggest that if knowledge of French wines and artisanal cheeses can be considered culture, then so also is knowledge of yam varieties and wild honeys. The discussion in Chap. 4 on the cultural significance of birds to the Solega also makes it clear that beliefs and practices well beyond the utilisation of an organism as food by the community in question need to be considered.

It is certainly not hard to find counterexamples to Berlin's claim in scientific taxonomy: potato, tomato, eggplant and nightshade are all scientific species under the genus *Solanum*, while dog, jackal, wolf, and coyote are all species of *Canis*—do English speakers really need "deliberate and conscious effort" to tell these organisms apart? Moreover, is a stoat (*Mustela erminea*) "almost instantaneously" distinguishable from a pine marten (*Martes martes*), but more easily confused with a polecat (*Mustela putorius*), simply because the latter is the same genus? I do not think Berlin's claim holds for folk taxonomies either—there is no compelling evidence from folk taxonomies that adult native speakers find it harder to distinguish "folk species" within a folk genus than to distinguish folk genera from each other. The only people in a community that this could be said of with some certainty would be young children and visiting ethnobiologists. In my experience, adults have provided me with erroneous subgeneric names as often as they have erroneous generic labels (see discussion on [140] below). Berlin's enthusiasm for Cain's work is also puzzling, given the latter's thoughts on the reality of the biological genus:

> A review of the interrelationships of different sorts of species and the evolutionary tree shows that the genus cannot now be regarded as a naturally discrete group either in relation to its ancestors and descendants, or at any one time. It is not necessarily definable by one single peculiar attribute, nor are its constituents monotypic, equivalent, essentially merely subdivisions of it, or themselves wholly discrete. It is monophyletic, but purely positional in rank, and a collection of phyletic lines, not an entity subdivisible into species. ([139], p. 108)

In addition to assuming that folk generics are universally more salient, both within and across languages, Berlin draws an explicit connection between folk and scientific taxa at this level of classification.

> For the higher vascular plants and larger vertebrate animals, generic taxa often approximate in their content the genera and species of Western scientific biology ([9], p. 25).

> Ethnobiologists will readily agree that in all systems of ethnobiological classification one can discover named groupings of plants and animals that represent Bartlett's "smallest groupings requiring a distinctive name" or, in Cain's rephrasing of Bartlett, "the smallest 'kind[s]' of plants and animals that can be recognised without close study." These fundamental taxa correspond… to taxa known as 'genera' in modern biological systematics ([9], p. 64)

However, Berlin is careful to note that the match between scientific and folk categories is not perfect:

> …folk genera do not in most cases correspond *perfectly* with taxa recognised as genera in the Western scientific system. Furthermore, subgeneric taxa… also do not generally correspond in a perfectly predictable way with taxa recognized as 'species' in Western taxonomy (p. 64).

While the last three statements sound somewhat contradictory, let us assume for the moment, that Berlin's position can be summed up as follows: For some groups of organisms, in particular higher vascular plants and large vertebrates, there is a good, but not perfect, correspondence between folk genera on the one hand, and biological genera on the other. Unfortunately, several counterexamples can be found even with this rather generous reading of Berlin's prediction—for many prominent, named groups of plants and animals, the correspondence between folk and scientific classification is not merely imperfect, but completely nonexistent. In Australia, where the biological landscape is dominated by a few very large genera like *Eucalyptus* and *Macropus*, folk generic names for the plants and animals belonging to these taxa in local indigenous languages almost invariably correspond to biological species within *Eucalyptus* and *Macropus* [13]. Note that this in sharp contrast with the Australian English folk terms used to describe these species—e.g. scribbly gum, red gum (kinds of *Eucalyptus*) and agile wallaby, tammar wallaby (kinds of *Macropus*), which are more in keeping with Berlin's predictions. Similar patterns emerge in the Amerindian nomenclature of oaks (*Quercus* spp.), where mononomial folk genera correspond to individual species, and there may or may not be a labelled superordinate category that corresponds to the English 'oak' [141]. As mentioned above, other large plant genera such as the Solanaceae will often be labelled by folk generics that correspond to scientific species, while labels such as the English 'monkey', 'deer', 'seal', or, for that matter, 'elephant', label taxa that include more than one scientific genus. Berlin's original 'General Principles', and their explication in the text of the 1992 monograph are flexible enough to accommodate such exceptions. However, the promotion of these 'General Principles' to the status of 'universals' [35] seems unwarranted, in light of the linguistic counter-examples described above.

At this point, it is useful to consider another of Berlin's generalisations that links scientific and folk genera:

> Focusing solely on monotypic genera as the most likely candidates for linguistic recognition, the following hypothesis is suggested:
>
> 1. If a scientific genus, x, is monotypic, it is highly likely to be given a distinct folk generic name.
> 2. The generic name will be restricted in its range of application to the single monotypic genus, x.

Berlin suggests a uni-directional effect which allows the prediction of aspects of folk taxonomy from scientific taxonomy. Here, Berlin fails to clarify if he is referring to local monotypy or monotypy on a global taxonomic level, as these two concepts are hugely different [142]. Moreover, if (1) scientific and folk genera often align closely (though not perfectly), (2) monotypic scientific genera are often labelled by exclusive folk genera, and (3) monotypic genera predominate in not only modern scientific taxonomy but also folk classifications (i.e. the hollow curve), it follows logically that the vast majority of folk generics should correspond to monotypic scientific genera. The preceding discussion has already demonstrated that this is definitely not the case for much of the named Australian flora and fauna. Waddy's [26] comprehensive inventory of Anindilyakwa (a northern Australian language) ethnotaxonomy clearly demonstrates a complete lack of binomial names—monotypic scientific genera abound in this part of the world, but they are treated no differently from polytypic genera by Anindilyakwa speakers, and practically every terminal taxon is a "distinct folk generic name". Berlin's perception-based model would imply that Anindalyakwa speakers are unable to perceive any strong similarities between, *e.g.* the nine different kinds of stingray that are named by mononomials (there is no superordinate 'stingray' taxon either). Waddy gets around this problem by the wholesale use of 'covert complexes', but this has the appearance of a solution to a problem that exists only in the analyst's mind. The discussion in Sect. 2.4.2 regarding the objectivity and consistency of morphology-based scientific taxonomy also calls into question the significance of any observed similarities of folk and scientific taxa, with regard to monotypy.

The existence of classificatory 'ranks' is key to Berlin's conception of the folk genus. The generic rank of a folk classification is meant to be the most salient to speakers of that language, partly because linguistic consultants provide 'generic' terms when asked by someone (usually the ethnobiologist) to name a plant or animal. However, it is worth asking oneself whether an interview situation (formal or informal), where someone from outside the speech community interviews a native speaker about plant names is really a natural speech situation that the latter would encounter often in his/her lifetime. Actual naturalistic language data are rarely, if ever, presented in ethnobiological studies, and it is therefore impossible to base claims of salience on anything other than the investigator's own subjective impressions. The few exceptions include Baker [13], who argues that in the Australian languages Ngalakan, Wubuy and Arrente, binomial compounds can be constructed not to refer to specific level taxa, but only to describe individuals of a species (along the lines of big dog, or black cat). The great majority of taxon names (even the names of polytypic scientific genera) in these languages are unanalysable monomorphemic labels. McKnight [77], on the other hand, demonstrates that the common way of announcing an upcoming hunt for turtle or dugong in Lardil is *ngada kendabaliyathu*, using the superordinate category *kendabal* 'dugong/turtle', instead of the generic terms *dilmirrur* 'dugong' and *bararun* 'turtle'. Here, the superordinate term is clearly more salient in the context of a particular kind of announcement (e.g. about a hunt from a canoe, using a spear).

Another piece of evidence that Berlin forwards to support the primacy of the generic rank is the child language acquisition study by Stross [140], which shows that Tzeltal children (this cross-sectional study involved children around the age of four) learned the generic names of plants and animals before the specific and varietal names ([9] p. 63). The results of the Stross study were verified by a follow-up study by Zarger and Stepp [143]. However, recent child development studies have shown that even pre-verbal infants are able to form global categories (i.e. categories equivalent to 'animal' versus 'furniture') well before they are able to form a basic-level category (i.e. 'cat' versus 'elephant') [144]. This finding has been replicated in a longitudinal study by Pauen [145]. The results of these experiments clearly show that the developmental trajectory of pre-verbal infants goes moves from global to basic level categorisation, and not the other way round. If Stross' data, based on interviewing 4-year olds, is meant to demonstrate the primacy of generic over specific and varietal taxa, then surely Quinn and Johnson's results on 2-month-olds would strongly suggest that higher-order taxa are more salient than generic taxa.

Berlin's concept of rank is closely linked to his rules of ethnobiological nomenclature. Points 10 and 11 (see above) of his model require taxa of certain rank to have names which possess certain properties, while point 5 makes a strong claim regarding the conceptualisation and content of taxa of the same rank. As noted earlier, Berlin also places great significance on the scientific concept of the genus, asserting that genera tend to be perceptually salient, and that monotypic genera, in particular, played a central role in ethnobiological classifications. This is surprising, for although the concept of rank is deeply embedded in modern systematics, professional taxonomists would be extremely wary of claiming that members of a particular rank were somehow equivalent across taxa. Stevens [142, 146] has the following to say on the topic of hollow curves:

> …for such curves to represent anything about biology, the units should be comparable; there must be a rank of genus (for example), all members of which have the same properties. Biological classifications cannot readily be compared in this way, even within flowering plants, even within a family. Antoine-Laurent de Jussieu, Bentham and Hooker, and Charles Bessey, to name just three prominent botanical examples, made classifications that, when interpreted as hierarchies, cannot be converted directly even to the authors' own ideas of relationships.

In any case, the Australian languages discussed above clearly violate Berlin's rules, as "primary names" are used to label both plants and animals right through the folk taxonomy. The Lardil taxon *bararun* 'turtle', for instance, which might be considered a folk generic due to its inclusion in the superordinate *kendabal* 'dugong/turtle', contains the subordinate taxa *yuburu* 'leatherback turtle', *lebulbul* 'hawksbill turtle' and *dunkumudin* 'greenback turtle', which bear no resemblance to each other, or to their category label *bararun*. An analysis of the familiar English folk taxon 'dog' is also illuminating in this respect. In Berlin's scheme, labels such as 'dog', 'cat' and 'horse' would be classed as folk generics, by virtue of their primary names, and the fact that they label perceptually salient 'natural kinds'. While the subordinate groups 'sheep dog', and 'bulldog' behave like good subgeneric (specific) taxa, in that they possess secondary names which make reference to the superordinate taxon, the vast majority of dog breeds are labelled by primary names, such as 'terrier', 'beagle', 'mastiff' and 'boxer'. Of course, some of these specific taxa can be further subdivided into 'varietal' taxa, such as 'fox terrier', but the specific taxa themselves are mostly primary names. Berlin's Point 11 makes provision for culturally significant folk species-level organisms to be named by primary names, but is a 'great Dane' (complex primary name) really more culturally significant than a 'sheep dog' (secondary name)? Furthermore, the only way to resolve the even more serious objections, that can be raised when comparing taxa of the same rank from widely separated groups, is to put forward explanations that are so *ad hoc* as to be completely meaningless. Could one ever convincingly argue that 'rainbow trout', 'Dover sole' and 'red mullet' have had secondary specific names thrust upon them because they are somehow less culturally salient than 'beagle'? Or that a 'Rhode Island Red', with its complex primary name is more highly valued, culturally speaking, than a 'fox terrier'?

In the following section, I attempt to demonstrate that at a very general level, Solega shows significant departures from the predictions made by Berlin's model. More specific instances of departures are presented in Chaps. 3 and 4.

2.6.1 Solega

Evidence from the Solega language strongly supports the hypothesis that utilitarian/cultural concerns largely determine which organisms in the natural environment are to be named. This is particularly true of plant names, which is by far the richest semantic domain in the language, in terms of total lexeme count.[5] Practically all the named plant species in the language have a use in Solega culture, from the tallest rainforest trees that provide wood, edible fruit or honeybee nesting sites, to the most unassuming prostrate herbs, whose leaves can be eaten. One significant group of plants whose members have not been named is the orchids. This is at first surprising, because the members of this large and diverse group are visually very striking (Fig. 2.1), and the Solega are

Fig. 2.1 All orchids are called *baccaṇike* in Solega. No subordinate names exist in the language. Photos by the author

[5] I have been able to document well over 500 plant names to date, far more than the 80 or so bird names, and the 40 or so names of other animals.

certainly aware of their presence on trees, or in the case of ground orchids, among the forest understorey. However, the Solega have no use for these plants, and as a result, their language has no name for these species. The group as a whole is referred to by the terms *baccaṇike* or *mara banja* 'orchid', presumably a generic-level term in Berlin's scheme. No other specific-level terms exist in the language.

This situation contrasts sharply with the Solega ethnotaxonomy of *aṇabe* 'mushroom', which happens to be the superordinate term for a handful of lower-level taxa. One common feature that links these named subordinate taxa is that they are all edible mushrooms that the Solega harvest from the forest during the rainy season (Fig. 2.2). Once again, the Solega are well aware of the existence of many other types of mushroom that also sprout in the rainy season, but when asked to provide names for these species, the reply is often a simple *na:vu adava tinnalle* 'we don't eat those', indicating a clear link between edible status and named status.

One of my primary consultants, M. R. Madha, was given a digital still camera with which to record any mushroom types that I may have missed. On a subsequent visit to the field site, I was shown dozens of photos of morphologically dissimilar mushroom species, each introduced with the statement *i:ga i: vicitrava:da aṇabeya no:ḍi* 'Now look at this interesting mushroom!' Madha, who had been so successful in locating and photographing these mushrooms, was, however, unable to name any of them, primarily because they were all considered inedible (Fig. 2.2a–h). Even the visually stunning 'dancing girl' mushroom (*Phallus indusiatus*), so named by visiting biologists (Fig. 2.2d), has no name in Solega, although practically everyone had seen it growing in the forest at one time or another.

On one of our many walks along forest trails, Madha mentioned a plant called *sã:yã* (sometimes also called *sã:yegã*), which at the time was new to me. When asked to describe this supposedly common plant, Madha replied, *adu road pakkadalli ja:sti irutte, ondu cikka giḍa, hu: illa, haṇṇu illa, ye:nu upyo:ga illa* 'it grows in large quantities alongside roads, it's a small plant without flowers or fruits, we have no use for it'. Madha was free to describe the plant (a fern, in this case) in any way he wished, but instead of mentioning possible features like the finely divided leaves, or the special way in which the leaves unfurl, he chose to mention the fact that it was a plant that was, in effect, 'useless'. As in the case of San Juan Gbëë Zapotec [147], fern names are highly under-differentiated in Solega, and, in fact, all ferns are called *sa:yã* or *sa:yegã*, despite highly obvious differences in leaf morphology. Labels such as *aḷḷa sa:yã* 'river fern' or *arre sa:yã* 'rock fern' may sometimes be offered as names for particular specimens, but it seems to me that these are used only for descriptive purposes. Figure 2.3 shows two very different ferns that were growing on the same hill, but they were both identified as *arre sa:yã*, simply because they were growing on a rocky substrate.

Occasionally a Solega consultant and I would come across a tree that was not endemic to the local forest. In many cases, these were plants that had been imported to the region from the lowlands by non-Solega migrants—plants that are otherwise common throughout India, such as the Gulmohar (*Delonix regia*), as it is known in the Hindi-speaking parts of the country. As expected, my guide would reply that there was no Solega name for that plant, and explain that it was a *siṭi giḍa* 'city

Fig. 2.2 (**a–j**) Inedible fungi are all labeled *uccaṇabe*. (**k–l**) edible mushrooms have binominal labels: (**k**) *eṇṇe aṇabe*, (**l**) *ko:ḷi aṇabe*. Photos by M. R. Madegowda

Fig. 2.3 Two morphologically very different types of fern that grow on a rocky substrate are both called *arre aṇṇe* 'rock fern'. Photos by the author

plant' or *so: giḍa*. I eventually realised that *so:* was the English loan word 'show'. My guide's intention was to inform me that such plants, that had been brought from the city, were only 'for show', and were otherwise quite useless.

For countless amateur (and professional) Western naturalists, the collection and identification of butterflies and moths has been a staple preoccupation—perhaps even an obsession—for well over two centuries. As a result, vernacular English names abound for the many thousands of lepidopteran species discovered around the world to date. The many butterflies that can be found in the Solega forest, however, live out their lives in complete anonymity, kept company only by their nameless host orchids and ferns. All butterflies and moths are simply called *siṭṭe* in Solega, despite the considerable morphological diversity present in this group (Fig. 2.4a–m). Insects of all shapes, sizes and colours can be seen flapping around understory plants in the spring months, ranging from dull brown moths to gaudy, metallic butterflies with wing spans equivalent to those of small birds (many of which, incidentally, are also called *siṭṭe*). These insects appear in large numbers every year, but as the Solega have no use for them, the taxon *siṭṭe* contains no subordinate terms. Even the highly distinctive *Macroglossum stellaratum*, which is so unusual for a lepidopteran that it has been given the English name Hummingbird Hawkmoth, is left out of the Solega lexicon, my respondents saying only that it was *ondu tarada huḷa* 'a kind of bug' (Fig. 2.4n). Solega has no name for even large solitary bees and wasps that are of no practical use (Fig. 2.4o) but at least four kinds of *je:nu* 'honeybee', which are colonial and yield different kinds of honey, are named (Fig. 2.5). Interestingly, a nearly identical situation has been reported by Hunn [62], for the North American language Sahaptin.

Fig. 2.4 (**a–m**) Butterflies and moths are all called *siṭṭe* in Solega. (**n–o**) The hummingbird hawk-moth (**n**) and solitary bees (**o**) have no specific names. Photos by the author

Fig. 2.5 Four species of honeybee are named in Solega. (**a**) *hejje:nu*, (**b**) *thuḍuve je:nu*, (**c**) *kaḍḍi je:nu*, (**d**) *nesari je:nu* (*white arrowheads* indicate the heads of two individuals). Photos by the author. Previously published in [148]

Table 2.1 indicates that plants—including trees, vines, herbaceous plants and grasses—mostly have two-part (and sometimes three-part) names, with the final slot of the compound being occupied by the superordinate label meaning 'tree', 'vine', 'plant' or 'grass'. This mirrors the San Juan Gbëë Zapotec naming system described by Hunn [147]. As discussed further in Chap. 4, the same tendency exists, but to a weaker extent, in the case of bird names, and is altogether absent in the names of mammals. However, the named plants in the Solega lexicon far outnumber the named mammals and birds combined, and it can be safely concluded that the folk generics in this language are mostly *not* mononomials. Moreover, different nomenclatural strategies exist in Solega for referring to different types of organisms.

Table 2.1 Solega nomenclatural patterns across various organism types

	Obligate secondary	Optional secondary	Obligate primary
mara 'tree'	*a:guri mara* *ta:re mara* etc.	None recorded	None recorded
ambu 'vine'	*nẽvẽ ambu* *bella:re ambu* etc.	None recorded	None recorded
giḍa 'herb'	*oṇṭe giḍa* *kaṭṭu:ḷi giḍa* etc.	*phaṇaki (giḍa)* *mu:rele na:ga (giḍa)* *etc.*	None recorded
hullu 'grass'	*jabbar hullu* *ba:ṇe hullu* etc.	None recorded	None recorded
hakki 'bird'	*araḷakki* *maṇṇakki* etc.	*uṇṇigirika(n-akki)* *karaḷi(y-akki)* etc.	*ko:ḷi* *natta:re* etc.
pera:ṇi 'mammal'	None recorded	None recorded	*aḷḷugã* *suṇḍa* etc.

2.7 Conclusion

In terms of folk ethnobiological nomenclature, the data presented here demonstrate, for instance, that the ethnobiological lexicon of languages can be dominated by binomials or secondary lexemes, even at the 'generic' level (Solega), or may have only unanalysable mononomials at the 'subgeneric' levels of their taxonomy (many Australian languages). Other researchers have reported that ethnobiological classifications may be organized in ways other than inclusive hierarchies (e.g. [149] for Sahaptin) or that Berlin's nomenclatural rules are better applied to plants than to animals ([150] for Ndumba). Still others have suggested that some societies may completely lack a 'standardised' ethnotaxonomy, but may instead tolerate a considerable degree of dissonance, leading to a taxonomy that is "*inherently dynamic and subject to negotiation*" ([60] for Wola).

The existence of strong similarities between scientific taxonomy and folk taxonomy is meant to be a crucial piece of evidence supporting Berlin's claim that ethnobiological classifications are primarily shaped by the cognitive similarities shared by all humans, with cultural or utilitarian factors playing a very small role. Directly addressing this point, the taxonomist P. F. Stevens, in reviewing Berlin's 1992 publication, expressed surprise that the latter seemed to take the veracity of scientific classifications for granted, when they were clearly "*as much mirage as etic grid*" ([142], p. 294). Furthermore, although Berlin insists that Ralph Bulmer's 1974 paper [151] on the PNG language Kalam "*provided the first independent corroboration of the striking similarities among ethnobiological systems of classification*", Bulmer's own thoughts on the universality of certain concepts presented in e.g. [12] were less conciliatory:

> … I do not believe that we can yet demonstrate correspondences between the cognitive statuses of folk taxa and the nomenclature applied to these which are of sufficient intra- and cross-cultural regularity to enable us to arrive at a simple typology. There is an obvious danger in advancing, prematurely, a typology of the kind Dr. Berlin proposes, namely that it may lead ethnographers and lexicographers to distort data by forcing it into inappropriate pigeonholes, and in particular into failing to appreciate and record the degree of flexibility and elasticity which is probably a very general feature of folk taxonomies.

It would be foolish to try and argue that human cognition played little or no role in determining the architecture of ethnobiological classifications. However, an unbiased appraisal of many, if not all, languages should reveal that cultural/utilitarian factors also have a significant hand in determining not only what is named, but also how it is to be named. As Bentley and Rodriguez [152] report on the naming of insects in Honduras, cultural importance and cognitive constraints jointly influence folk nomenclature. Such an idea is not new: according to Hays [153], folk classifications are "*products of a number of complex, interacting factors: biological discontinuities in nature, chance historical events, "utilitarian" human concerns, human cultural concerns in a broader sense, intellectual curiosity, and constraints deriving from the nature of human perception and cognition*". It is entirely likely that many of Berlin's rules do indeed operate in many languages, perhaps the majority, but that simply makes them 'tendencies', and not 'universals'. The fact that several unrelated languages behave in the same way cannot be regarded as conclusive—many unrelated languages use a base 10 numerical system, and indeed, the existence of 10 fingers on all humans (a 'perceptual universal') may have contributed to the commonness of such systems [154, 155]. However, genuine non-base-10 numerical systems, including those utilizing tallies on body parts other than the fingers, are frequently reported by fieldworkers [156, 157]. Berlin has been criticised for producing a "*universal model designed to accommodate all data; a model, into which each analyst was expected to fit his or her data*" ([14], p. 13), a model, which has the aesthetic and intellectual appeal of making the data organized within its frame look more coherent and tidy. The science of biology—also called the 'science of exceptions' [158])—is at least as messy as the study of human languages and cultures, but biologists are always careful to make room for exceptions. Perhaps, as Kay and Regier [111] suggest, it is time to "*move the field beyond a familiar rhetoric of 'nature versus nurture', or 'universals versus relativity', to new concepts and new questions*".

Chapter 3
Plants in Solega Language and Culture

3.1 Introduction

The location of the Biligiri Rangaswamy Hills at the confluence of two important mountain ranges of south India and the existence of numerous habitat types within these Hills have conspired to bestow upon the forest home of the Solega an abundance of plant life. Solega *po:ḍu* 'hamlets' are scattered all over the B. R. Hills, and people are well acquainted with habitats that biologists would call 'lowland scrub', 'moist deciduous', 'dry deciduous', 'evergreen' and so on. Unsurprisingly, the Solega language has its own words for such categories—although indigenous conceptualisations do not always match those of the biologist—and these and other landscape terms are described in more detail in Chap. 5. The Solega botanical lexicon consequently includes the names of lowland, drought-tolerant, thorny plants such as *Opuntia* sp. and *Ziziphus* spp. as well as large, montane, evergreen trees such as *Canarium strictum* and *Aphanamixis polystachya*. Knowledge of plant names and uses is not evenly distributed across the population, and people living in highland hamlets are, naturally, more at ease with identifying and naming highland plants.

To say that the Solega regard the forests of the B. R. Hills as their home would be an understatement. To older Solega, living in the forest is the natural state of affairs, and most would consider it an ordeal to face the heat, dust, noise and crowds of treeless lowland towns and villages for extended periods of time. The words of the late Heddini Basavegowda eloquently summarise the idealised Solega existence: *"Aḍaviliye baddu, aḍaviliye sa:yadu"* '(We were meant to) grow up in the wild, (we were meant to) die in the wild.' Congruous with this strong attachment to the forest are a number of other beliefs often expressed by older people: the ability of traditional forest foods to keep illnesses at bay, the superiority of natural remedies over the pills and injections dispensed by city-trained medical practitioners, and the need to regularly burn patches of forest not only for agriculture, but also to promote regrowth.

A. Si, *The Traditional Ecological Knowledge of the Solega*, Ethnobiology,
DOI 10.1007/978-3-319-24681-9_3

Ka:ḍu, a commonly used Solega word, is most readily translated into English as 'forest', but the way this word is used in everyday speech reveals its threefold significance:

1. *ka:ḍu* refers to the Solega homelands, as opposed to the territory (*na:ḍu*) inhabited by non-Solega people
2. *ka:ḍu* denotes various named forest or vegetation types (senses 1 and 2 are discussed in detail in Chap. 5)
3. *ka:ḍu* denotes the forested area beyond a settlement and its fields

It is usually in the context of the third sense above that people talk about the hundreds of plant species that are used for food, medicine, construction and the assembly of useful artefacts. Approximately 500 Solega plant names have been documented, of which around 300 have been identified botanically so far. While the uses of some of these plants are described later in this chapter, the first few sections deal with issues of plant classification and nomenclature and with the significance of plants and plant names in Solega culture.

3.2 Solega Attitudes Towards Plants

It is difficult to work in the B. R. Hills for an extended period of time without coming to the conclusion that Solega people are very fond of the plants growing around them. One often hears adult men talking about a tree near their village that is currently in bloom, or admiring the beauty of a rare flower. While being given a tour of a hamlet on a first visit, guests are shown not only sacred sites, water sources, fields and rock platforms (sites of activities such as grain-drying, laundry and machete-sharpening), but also large trees in or around the hamlet that have recently fallen over. In such cases, the visitor gets a clear sense of how important such a tree was—having stood at that spot probably all through the living memory of the hamlet's residents—and how, after the tree's demise, its life continues to be quietly commemorated.

As noted in Chap. 2, the orchids are a group of plants that the Solega do not use in any way, and do not distinguish at the species level. While clearly able to tell different orchid flowers apart visually, as well as appreciate their beauty, Solega people lump all orchids together under the category *baccaṇike*. Their mild disinterest towards this diverse and often conspicuous group of plants is matched by the mild bemusement with which they regard the obsessive fervour shown by non-Solega (including the author, tourists, botanists and ecologists) in attempting to locate as many orchids as possible during visits to the forest. Solega guides accompanying such orchid-tragics will nevertheless point out new orchid species in a helpful and good-natured way, saying, "*No:ḍu, alli innondu a:rkiḍ ade!*" "Look, there's another orchid!" often using the English word *a:rkiḍ* 'orchid' in preference over *baccaṇike*.

Not all plants are looked upon benignly, and the invasive *Lantana camara* is a notable exception. This plant is blamed not only for driving several understorey plant species to local extinction, and preventing tree saplings from sprouting, but universally condemned for making travel through the forest on traditional paths difficult. Travelling through the forest is also made more dangerous, as dense thickets of *Lantana* can be tall enough to hide an elephant, causing even normally cautious Solega people to wander too close to these animals. This plant is usually referred to as *ro:jiga* (possibly from the English 'rose'), but it can also be called *uṇṇi gutti* 'tick bush', although it is unclear whether this is a reference to the fact that these bushes can harbour ticks, or whether being scratched by the thorns of the plant is being likened to tick bites.

3.2.1 Plants Used to Be People

There are many instances to be found in Solega folklore that show how Solega people identify with a range of non-human organisms—examples can be found, in Solega mythology, of organisms that are said to possess human-like traits. In some cases, this is the result of an organism once having been a human, who was forced to take on a non-human form due to extenuating circumstances. Trees were once humans who had avoided offering any aid to the god *ma:desurã* during his battle with the demon *savaṇa*. Seeing that these same people expected to profit from *savaṇa*'s death in spite of their inaction, *ma:desurã*, in a fit of rage, cursed his cowardly followers, and turned them into trees where they stood.

Perhaps as a direct result of this myth, Solega people rarely cut down entire trees, preferring instead to remove individual branches for firewood or to harvest the fruit growing on them. It is taken for granted that a *su:lu* 'soul' resides in all plants and animals, and a tree that is cut makes a mournful *gi:nk* sound as its *su:lu* departs (this sound has also been characterised as *a: a: gri: gri:* in [86]). To lessen the spiritual trauma being caused to the tree, the woodcutter places a special rock next to it, into which the *su:lu* enters. This is reminiscent of the *kallugombe* that form a vital part of Solega funerary rituals—these are smooth rocks that represent the soul of a deceased person, which are placed, along with older *kallugombe*, in clan-specific sacred sites in the forest.

3.2.2 Relationships with Individual Trees

As mentioned above, individual trees can come to hold special significance for Solega people. Sometimes, trees that are afforded a high status for cultural reasons can even be given proper names. These proper names often live on to serve as place names long after the tree has died or fallen over, thus preserving the memory of the individual tree, the ethnospecies it belonged to, and any special properties associated with it. Trees that are singled out for naming are almost always large, and possess one of a small number of key attributes. Chief among these is being a host to large numbers of *hejje:nu* (*Apis dorsata*) beehives every year; namely, that it be a *je:nu mara* 'bee tree' (this phenomenon is discussed in the context of Solega honeybee knowledge in Chap. 7). Trees can also be named for being sacred, for possessing unusual morphological characteristics, or for being associated with a dangerous animal, like a tiger (Sect. 3.8.4).

3.3 General Naming Conventions

Very broad patterns in the naming of plants and animals are first presented in this section. The reader will notice that while a small number of naming conventions are common to all plant and animal groups, key differences do exist in the naming of categories such as trees, birds and mammals. The differences are systematic, and they make it difficult to formulate a general, unified 'Solega nomenclature' that might apply to the members of all ethnotaxa.

3.3.1 The Prefix *he* □-

This prefix plays a limited, but significant, role of singling out a particular organism as the largest of a contrast set. Its role is limited, because it is not a very productive prefix, having so far been documented in only a handful of words. These include *hejje:nu* 'giant honeybee' (*Apis dorsata*) from *je:nu* 'bee', *hebbidiru* 'thorny bamboo' (*Bambusa arundinacea*) from *bidiru* 'bamboo', *hebbe:vu* 'Indian Mahogany' (*Toona ciliata*) from *be:vu 'Azidirachta indica', hebbaddu* '?vulture' from *haddu* 'raptor' and possibly *heggaṇa* 'a kind of shrew' (although there appears to be no **kaṇa* or **gaṇa* in Solega as it is currently spoken). Rarely, this prefix is also used for inanimate objects, as in the case of *heggave* 'king beam' (i.e. the two load bearing posts, or *kave*, of a construction that hold up the roof).

3.3.2 The Epithet *uccu*

This epithet *uccu* is only used in the botanical domain, to denote plants (and fungi) that are of no utility to humans. Literally meaning 'mad' or 'crazy' (as in the phrase *uccu mansa* 'crazy person'), it can label inedible mushrooms (*uccaṇabe*), a non-native, non-utilised species of *Indigofera* (*uccu maruḷi*), and a handful of other non-utilised plants, such as *uccu eḷḷu* and *uccu tagate*—the latter two have yet to be identified scientifically. Interestingly, this pattern also extends into the meteorological domain, as 'heavy rain with large raindrops that can flatten crops' is called *uccu maḷe* 'crazy rain'.

3.4 Patterns in Solega Plant Classification

As mentioned above, the majority of Solega plant names are two- or three-part compounds, making them secondary lexemes (*sensu* [9]), much in the same way as described by Hunn for Zapotec [147]. Moreover, the final element is almost always one of four superordinate category labels that indicate whether the plant in question is a tree, vine, herb or grass. There are some exceptions to this rule, which normally manifest themselves when certain edible plants are being talked about. In spontaneous speech (as opposed to a formal elicitation of plant names), yams, edible greens and some medicinal roots and corms are referred to by alternative terms, such as *geṇasu*, *soppu* and *be:ru/geḍḍe* respectively. The first of the following examples, *nu:re geṇasu*, is a commonly eaten yam, that grows deep underground, and has a long, vertically oriented tuber. *Geṇasu* is the Solega word for all edible underground tubers and corms (except for the exotic potato), and this culinary label is often used to refer to the whole plant, whose name, strictly speaking, should be *nu:re ambu* '*nu:re* vine', in reference to its growth habit.

nu:re geṇasu/ambu
nela go:ṇi soppu/giḍa
a:sa:ḍi be:ru/ambu

The first element of a two-part name is usually unanalysable, whereas in the case of three-part names, the first element is often a descriptive epithet. Possible words that can be found in this position include those that describe overall size (*doḍḍa* 'big', *saṇṇa* 'small'), habitat or substrate (*ka:ḍu* 'forest', *ka:nu* evergreen forest', *beṭṭa* 'mountain', *na:ḍu* 'lowland', *kallu* 'rock', *aḷḷa* 'stream', *arre* 'rock platform', *nela* 'ground', *ni:ru* 'water'), taste (*kaĩye* 'bitter'), anatomy (*muḷḷu* 'thorny') and supernatural associations (*daiya* 'witch'). The *uccu* 'mad' epithet mentioned above also belongs to this class. Plants that belong to the three-part name category are almost always grouped (and contrasted) with at least one other plant that either has a different epithet as a first element, or lack an epithet altogether. Thus, *kaĩye suṇḍe giḍa* 'bitter *suṇḍe* plant' is said to belong to the group that also includes *muḷḷu suṇḍe giḍa* 'thorny *suṇḍe* plant' and *suṇḍe giḍa* '*suṇḍe* plant'.

Two Part Names:

kakke mara	*Cassia fistula*
poṭle ambu	*Passiflora subpeltata*
iṇḍãna giḍa	*Pterolobium hexapetalum*

Three Part Names:

kaĩye suṇḍe giḍa	'bitter *suṇḍe* plant'
beṭṭa ko:li mara	'mountain *ko:li* tree'
doḍḍa puḷḷuḷi giḍa	'big *puḷḷuḷi* plant'

A third category of plant names consist of noun or verb phrases that describe a salient feature of the plant or its medicinal (or other) uses. Some of the references made by the examples given below are quite obvious. Thus, *hoṭṭeno:vina ambu* 'stomach-ache vine' is a plant whose root is used to treat stomach ache, while *mu:r ele na:gada giḍa* 'three-leaf *na:ga* plant' large, trifoliate leaves. The wild passion fruit *ka:ge ma:ri ambu* 'crow chick vine' is so named because of the appearance of its edible, ripe fruit, while the small, weedy *Tridax procumbens* is called *o:le* 'ear-ring' in at least one Solega settlement (*Bu:ta:ni po:ḍu*) because the children there use the long flower stalks to make earrings. This plant also has the alternative names of *garugada giḍa* or *ḍa:bu giḍa* in other settlements.

Descriptive Names:

ha:lu so:rada geḍḍe	'milk releasing root'
hoṭṭeno:vina ambu	'stomach ache vine'
ko:ḷi kuṭumana giḍa	'cockscomb plant' *Gloriosa superba*
ka:ge ma:ri ambu	'crow chick vine' *Passiflora edulis*
o:le giḍa	'earring plant' *Tridax procumbens*
mu:r ele na:gada giḍa	'three leaf *na:ga* plant'

The patterns seen in the naming of plants can also be observed to a large extent in the bird names discussed in Chap. 4. Unlike the bird names, however, where the superordinate category label *hakki* 'bird' can often be omitted, the same cannot be said for plant names, except in a small number of linguistic contexts. These contexts are described below in Sect. 3.6. In the next Section, the issue of correspondences between Solega plant names and the scientific classification of plants is taken up.

3.5 Solega Ethnospecies and Scientific Classification

For the vast majority of plants, there is a one-to-one correspondence between a Solega ethnospecies and a botanical species. This applies to all three categories of names described in the previous section. In a handful of cases, one ethnospecies label can refer to a large number of scientific species; this is particularly true of plants that Solega people regard as having no utility to either themselves or to other animals. Included here are categories such as *baccaṇike* 'orchids', *sa:yã* 'ferns', *pa:se* 'lichen/moss' and *aṇṇe* 'succulents like *Bryophyllum* spp.' Naturally, these are open-ended categories, whose exact membership is of no relevance to Solega speakers.

In a very small number of cases, more than one ethnospecies maps onto a single biological species. For instance, *biḷi sampage* and *kenda sampage* label the white and orange varieties of the sacred *Michelia champaca* tree respectively, while *ma:vũ* and *tore ma:vũ* are the Solega names for the cultivated and wild varieties of the mango *Mangifera indica* respectively. While the one-to-many and many-to-one patterns are commonly found in ethnoclassification systems all over the world, the more interesting issues in Solega ethnotaxonomy lie in the relationships between Solega categories and biological genera. These relationships are discussed next, in the context of Berlin's [9] claims regarding the saliency of 'folk generics' and scientific genera.

As mentioned earlier, Solega plant naming conventions are at odds with one of Berlin's nomenclatural universals, namely that the vast majority of 'folk generic' names should be 'primary names'. Solega plant names require an obligatory superordinate category label (*mara* 'tree', *ambu* 'vine' and so on), and therefore do not behave like the 'folk generic' names predicted by Berlin's framework. Nevertheless, plant names such as *kakke mara* and *poṭle ambu* are indeed the 'folk generics' of the Solega ethnobotanical classification scheme, as they are the names people use in everyday situations to talk about the plants in their environment. A further claim made by Berlin, which will be investigated further here, is that there is usually a one-to-one correspondence between a folk generic and a scientific genus, and that this pattern is most evident among the names of flowering plants.

3.5.1 Plant Groupings in Solega

A number of Solega plant categories contain one or more subordinate members whose names bear some resemblance to the superordinate category label. Such groupings resemble the 'polytypic genera' described by Berlin, and exemplified by taxa such as *oak* (genus *Quercus*) > *white oak* (*Quercus alba*), *chestnut oak* (*Quercus prinus*), etc. A key feature of the *oak* example is the close relationship between the English folk taxa and corresponding botanical genus and species. Similar polytypic name complexes exist in Solega, but there is considerable variation in the way in which the members of such a complex map onto a scientific genus. Table 3.1 shows how some polytypic genera behave in accordance with Berlin's predictions (full match). Many Solega taxa map onto species of more than one Linnaean genus, and conversely, there are numerous instances where the species of a polytypic scientific genus are labelled by completely different, often monotypic Solega folk generic names. As an example of the former situation, the Solega plants *a:le, beppa:le* and *komba:le* (the latter two being analysable as types of *a:le*) correspond to not a single polytypic Linnaean genus, but to three genera, namely *Tylophora*, *Holarrhena* and *Wrightia* (and hence, show no match). An example of the latter situation is *Bauhinia purpurea* and *Bauhinia racemosa* being labelled by the completely unrelated Solega names *kancuva:ḷa* and *kitta:rsã* respectively (again, no match).

A simple quantitative analysis of the names in the above table shows that the number of groups of either nomenclatural system, which show absolutely no correspondence to the other nomenclatural system, far exceeds those that match fully. Figure 3.1 shows that almost two-thirds (63 %) of polytypic Solega ethnotaxa show no correspondence with Linnaean groupings, while the members of over half of the polytypic Linnaean generic groupings are labelled by completely unrelated Solega names.

Even if the values for 'partial match' were to be combined with 'full match', the total number of groups in the resulting category would only equal those in 'no match'. In other words, it would only be appropriate to conclude that groupings of Solega taxa can sometimes coincide with the groupings found among Linnaean species. The data presented here do not support the notion of any systematic regularity between Linnaean genera and folk genera.

Table 3.1 Correspondence between Solega plant names and Linnaean taxonomy, divided into three categories: groups that show no match, groups that show a partial match, and groups that match fully

No match		
Polytypic Solega ethnotaxa and their relation to scientific names		
Solega name	**Scientific ID**	**Alternative ID**
a:le	*Tylophora indica*	
beppa:le	*Holarrhena pubescens*	
komba:le	*Wrightia arborea*	
araḷu	*Ricinus communis*	
ka:naraḷu	*Melicope lunu-ankenda*	
koṭṭaraḷu	*Jatropha* sp. (*J. curcas, J. gossypiifolia*)	
hebbe:vu	*Toona ciliata*	
kaĩye be:vu	*Azadirachta indica*	
nela be:vu	*Andrographis paniculata*	
(na:ḍu) du:pa	*Soymida febrifuga*	
ka:nadu:pa	*Canarium strictum*	
seluge du:pa	*Boswellia serrata*	
ekka	*Calotropis gigantea*	*Asclepias gigantea*
ambu ekka	*Dregea volubilis*	*Asclepias volubilis*
biḷi ekka	*Gomphocarpus physocarpus*	*Asclepias physocarpa*
ka:ḍekka	*Rotheca serrata*	
kempu ekka	*Asclepias curassavica*	
je:neraku	*Mallotus tetracoccus*	
kaĩye eraku	*Clerodendrum infortunatum*	
hokke eraku	*Macaranga peltata*	
go:ṇi	*Ficus drupacea*	
arre go:ṇi	*Cyanotis adscendens*	
nela go:ṇi	*Bacopa monnieri*	
honne	*Pterocarpus marsupium*	
ka:rhonne		
sirhonne	*Bridelia retusa*	
nela honne	*Tephrosia calophylla*	
kaḷḷi	*Euphorbia* spp., such as *E. umbellata*	*Synadenium grantii*
muṇḍugaḷḷi/guṇḍugaḷḷi	*Caralluma umbellata*	
pa:pasagaḷḷi	*Opuntia* sp.	

(continued)

Table 3.1 (continued)

No match		
doḍḍa ka:re	*Psydrax dicoccos*	*Canthium dicoccum*
iṭṭu ka:re/na:ḍu (doḍḍa) ka:re	*Canthium coromandelicum*	
magga:re	*Catunaregam spinosa*	
tambiṭṭu ka:re	*Meyna laxiflora*	
aḷḷa ko:li	*Ludwigia octovalvis*	
beṭṭa ko:li	*Wendlandia thyrsoidea*	
biḷi ko:li	*Colebrookea oppositifolia*	
oṇṭe	*Bidens pilosa*	
doḍḍa oṇṭe	*Cosmos caudatus*	
kuri oṇṭe	*Urena lobata*	
doḍḍa kuri oṇṭe	*Triumfetta pilosa, Triumfetta rhomboidea*	
meṇasi:ge ambu	*Toddalia asiatica*	
meṇasi:ge mara	*Litsea josephi* (U)	
hora muni	*Biophytum reinwardtii*	
ola muni	*Mimosa pudica*	
te:ku	*Tectona grandis*	
nela te:ku	*Rotheca serrata*	
udi	*Radermachera xylocarpa*	
pe udi	*Oroxylum indicum*	
pa:dire	*Stereospermum tetragonum*	
na:ḍu pa:dire	*Dolichandrone falcata*	
anase	*Glochidion zeylanicum*	
ni:ranase	*Memecylon lushingtonii* (U)	
pa:paṭe	*Pavetta indica*	
ka:na pa:paṭe	*Chasallia ophioxyloides* (U)	
na:ḍu pa:paṭe	*Psychotria nigra*	
beṭṭa maruḷi	*Indigofera wightii*	
kallu maruḷi	*Schefflera venulosa*	

Polytypic scientific genera and their relation to Solega names		
Scientific ID	**Solega name**	**Alternative ID**
Albizia amara	ba:ge	
Albizia odoratissima	sele	

(continued)

Table 3.1 (continued)

No match		
Andrographis paniculata	nela be:vu	
Andrographis serpyllifolia	ka:sinasara	
Bauhinia purpurea	kancuva:ḷa	
Bauhinia racemosa	kitta:rsã	
Caesalpinia mimosoides	gi:jiga	
Caesalpinia decapetala	ba:ḍubaḷaka	
Clerodendrum chinense	maisu:ru mallige	
Rotheca serrata (included here because of synonym)	ka:ḍu ekka/nela te:ku	*Clerodendrum serratum*
Clerodendrum infortunatum	kaĩye eraku	
Curcuma longa	ka:ḍa arasu	
Curcuma sp.	suḷḷi	
Dalbergia lanceolaria	buluga	
Dalbergia latifolia	bi:ḍi	
Elaeocarpus serratus	bikkilu	
Elaeocarpus tuberculatus	kende	
Indigofera astragalina	uccu tagaṭe	
Indigofera sp.	beṭṭa maruḷi	
Litsea floribunda (U)	goṇḍe more	*Cylicodaphne floribunda*
Litsea josephii (U)	meṇasi:ge	
Mallotus phillippensis	ke:silu	
Mallotus tetracoccus	je:neraku	
Passiflora calcarata (U)	poṭle	
Passiflora edulis	ka:ge ma:ri	
Persea americana	beṇṇe	
Persea macrantha	karava:di	
Sterculia guttata (U)	ma:ṭe	*Astrodendrum malabaricum, Clompanus-malabarica*
Sterculia villosa	kallu sauwe	

(continued)

Table 3.1 (continued)

Partial match		
Polytypic Solega ethnotaxa and their relation to scientific names		
Solega name	**Scientific ID**	**Alternative ID**
kurrugu	*Strobilanthes* spp.	
gejje kurrugu	*Nilgirianthus heyneanus* (U)	
nela/doḍḍa kurrugu	*Nilgirianthus foliosus* (U)	? *Strobilanthes foliosa* (U)
nela kurrugu	*Justicia japonica*	
mallige/kanaka ambara	*Jasminum sambac*	
ja:ji mallige	*Jasminum azoricum*	
dore mallige	*Dahlia* sp.	
gersi mallige	*Hypericum mysurense*	
panni:ru mallige	*Portulaca grandiflora*	
su:lera/maisu:ru mallige	*Clerodendrum chinense*	
suttu mallige	*Tagetes erecta*	
aranelli	*Garuga pinnata*	
iṭṭu nelli	*Phyllanthus indofischeri*	
uḷi nelli	*Phyllanthus reticulatus*	
na:yĩ nelli	*Phyllanthus emblica*	
nela nelli	*Phyllanthus niruri*	
tumbe	*Leucas aspera*	
arre tumbe	*Leucas lavendulifolia* (U)	
beṭṭa tumbe	*Leucas eriostoma*	
nela tumbe	*Mitracarpus hirtus, Richardia scabra*	
matti	*Terminalia crenulata*	
ni:ru matti	*Terminalia arjuna*	
ambu matti	*Combretum ovalifolium*	
naravalu	*Olea paniculata*	
kaĩye naravalu	*Chionanthus ramiflorus*	
ka:na naravalu	*Chionanthus* sp.	

Polytypic scientific genera and their relation to Solega names		
Scientific ID	**Solega name**	**Alternative ID**
Acacia chundra	kaggali	
Acacia sp.	mukuḷi	
Acacia sinuata	biḷi si:ge	
Acacia leucophloea	na:yĩ bela/sujjilu	
Acacia pennata	ũsu si:ge	

(continued)

Table 3.1 (continued)

Partial match		
Canthium dicoccum	doḍḍa ka:re	
Canthium coromandelicum	na:ḍu doḍḍa ka:re	
Canthium travancorium	ambe	
Cassia fistula	kakke	
?*Cassia floribunda*$_1$, *Senna occidentalis*$_2$	tagaṭe	?*Senna floribunda*$_1$, *Cassia occidentalis*$_2$
Senna hirsuta	uccu tagaṭe	*Cassia hirsuta*
Chamaecrista mimosoides	seṇabbã	*Cassia mimosoides*
Senna auriculata	a:varike	*Cassia auriculata*
Senna montana	tagaṇi	*Cassia montana*
Grewia hirsuta	uḍupe, na:ḍu uḍupe, saṇṇa uḍupe	
Grewia tiliifolia	daḍasu	
Terminalia arjuna	ni:ru matti	
Terminalia bellirica	ta:re	
Terminalia chebula	araḷe	
Terminalia crenulata	matti	
Terminalia paniculata (U)	oluge	*Pentaptera* spp. (U)
Wrightia antidysenterica	beppa:le	
Wrightia tinctoria	doḍḍa a:luppe	
Wrightia arborea	komba:le	
Ziziphus rugosa/Z. xylopyrus	goṭṭi	
Ziziphus jujuba	na:ḍu goṭṭi	
Ziziphus oenoplia	soda:li	
Ficus nervosa	icci	
Ficus amplissima	biḷicci	
Ficus tinctoria	kallicci	
Ficus tsjahela	karicci	
Ficus benghalensis	a:la	
Ficus drupacea	go:ṇi	
?*Ficus hispida/F. exasperata*	sunkatti	
Ficus racemosa	atti	
Ficus religiosa	araḷi	

(continued)

Table 3.1 (continued)

Full match		
Polytypic Solega ethnotaxa and their relation to scientific names		
Solega name	**Scientific ID**	**Alternative ID**
goṭṭi	*Ziziphus rugosa*	
na:ḍa goṭṭi	*Ziziphus horrida* (U)	
icci/biḷicci	*Ficus amplissima*	
kallicci	*Ficus tinctoria*	
kari icci	*Ficus tsjahela*	
ju:jakki	*Lantana indica*	
biḷi ju:jakki	*Lantana veronicifolia*	
biḷi si:ge	*Acacia sinuata*	
ũ:su si:ge	*Acacia pennata*	*Senegalia pennata*
suṇḍe	*Solanum torvum*	
kaĩye suṇḍe	*Solanum violaceum*	
bu:du suṇḍe	*Solanum ferox* (U)	
ka:na suṇḍe	*Solanum erianthum*	
aṇṭuva:ḷa	*Sapindus longifolius* (U)	
na:ḍu aṇṭuva:ḷa	*Sapindus emarginatus*	
Polytypic scientific genera and their relation to Solega names		
Scientific ID	**Solega name**	**Alternative ID**
Asclepias curassavica	kempu ekka	
Gomphocarpus physocarpus	biḷi ekka	*Asclepias physocarpa*
Chionanthus ramiflorus	kaĩye naravalu	
Chionanthus sp.	ka:na naravalu	
Glochidion heyneanum	ka:nanase	
Glochidion zeylanicum	anase	
Nilgirianthus foliosus (U)	nela/doḍḍa kurrugu	*Strobilanthes foliosa* (U)
Nilgirianthus heyneanus (U)	gejje kurrugu	

Correspondences in both directions are shown: first in each category are Solega polytypic ethnotaxa and corresponding scientific names, followed by polytypic Linnaean genera and corresponding Solega names. All scientific names shown in the table are listed as 'Accepted' in www.theplantlist.org. Alternative IDs are scientific names that are listed as 'Synonyms' (and therefore, not the preferred names) in www.theplantlist.org, and are only shown in the table if they have a bearing on the match between the Solega and scientific names. (U) indicates an unresolved scientific name. The 'life form' terms *mara* 'tree', *giḍa* 'plant' and *ambu* 'vine' have been omitted for clarity

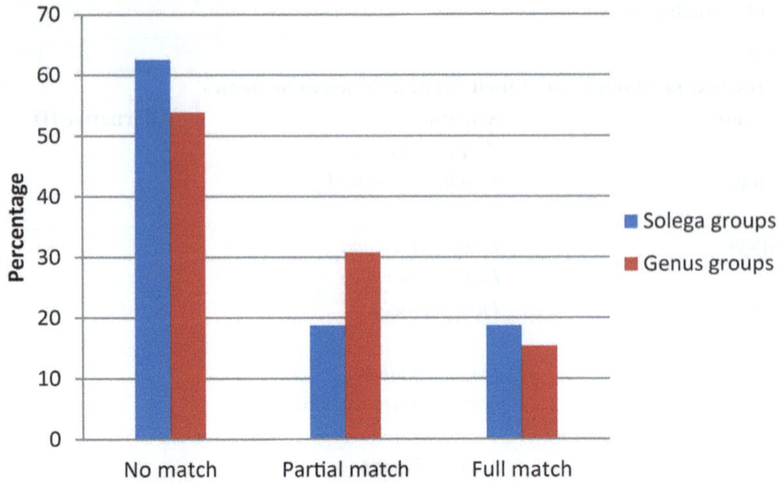

Fig. 3.1 Percentages of non-matching, partially matching and fully matching plant groupings (Solega ethnotaxa and Linnaean polytypic genera) from Table 3.1

3.6 Ethnospecies Names and Linguistic Conventions

While it is a useful and interesting exercise to analyse the elements of an ethnoclassification scheme in an abstract, decontextualized manner (as has been done above), it is important to complement such an analysis with some indication of how plant and animal names are actually used by the speakers of a language in naturalistic, everyday communicative events. Examining the use of a word within the context of a narrative or a conversation—even a word as seemingly straightforward as a plant name—can yield a much richer and more precise picture of the semantic range of such a word, and reveal potentially hidden semantic connections with other concepts and categories. The idea of firmly situating linguistic ethnobiological research within a social and cultural context is a central theme of this book, and the following paragraphs examine the use of ethnotaxonomic labels by Solega people in the context of naturalistic speech.

Chapter 6 describes how numerous trophic, ecological or temporal relationships between two or more biological entities, or between a biological entity and an abiotic feature of the environment can be encoded in Solega words. Plant names appear in compounds such as *beṇḍe hu:vina visa* '*Kydia calycina* flower poison' (a millet disease that coincides with the flowering of the *beṇḍe* tree) and *baise hu:vina jiḍi* '*baise* flower drizzle' (light rain that falls in December) and *maruḷi je:nu* '*Indigofera* sp. honey' (a particularly delicious type of honey made from ground-dwelling bees)—these are clear indications of the flowering of plants being used to mark important environmental events and phenomena.

The form of a plant or animal name can be modified depending on the type of utterance in which it is used. While the full, 'citation form' of a plant name invariably contains a superordinate label such as *mara* 'tree', this label can be optionally omitted in certain contexts. The most common type of context where this happens is the recitation of a list, such as when a person is asked to name the types of trees that one usually finds in a particular forest type. The following extract from a larger narrative in which the speaker describes the features of evergreen forests illustrates this phenomenon:

Example: *Ka:ninalli ha:le mara ade, thuruve mara ade, kakkilu, bikkilu, bella:ḍe, ku:ma: vũ, kende, soravilu ...*

In the evergreen forest, you find *Wrightia* sp., *Ligustrum perotteti, Celtis tetrandra, Elaeocarpus serratus, Neolitsea zeylanica, Cinnamomum sulphuratum, Elaeocarpus tuberculatus, Acrocarpus fraxinifolius...*

Plant names can also be modified depending on their location in an extended narrative. Names are often shortened when they are first presented at the start of a narrative to herald the start of a new topic of conversation. Typically, a person will first state the bare plant name in isolation at the start of his/her conversational turn, and may revert to using the full name later in the turn. This phenomenon has also been recorded in the use of bird names, where the superordinate label *hakki* 'bird' can be dropped from some names in such a situation (see Sect. 4.3 in Chap. 4).

Finally, large, culturally important trees that are given proper names (see Sect. 3.8 below) do not require the *mara* 'tree' label, and are referred to using the bare species name, with appropriate locative or descriptive epithets. Examples include *doḍḍa sampage* 'big *Michelia champaca*' and *gummana guḍḍe ta:ri* 'owl hill *Terminalia bellerica*', which are the names of an important sacred tree and a bee tree respectively (Table 3.3).

As with many other languages of the Indian subcontinent, the Solega language allows modification of a noun through a simple, regular lexical operation to increase the scope of the noun to include additional concepts that are semantically or contextually associated with it. This operation is carried out by reduplicating the target word, and modifying the initial sound of the second token. Different Indian languages require a different sound modification, and in Solega and Kannada, the first syllable of the reduplicated token is replaced with a *gi-*. As an example, the Solega word *a:ne* 'elephant' can be modified to *a:ne-gine* 'elephants and other similar creatures'. Although the phrase "and other animals" at first appears vague and unhelpful, Solega speakers have no difficulty in understanding such constructions, as the context of the utterance usually only affords one interpretation. In the following example, *a:ne-gine* is used in the context of the speaker's fields being raided by wild animals. This context narrows down the potential types of animals being referred to by the speaker to those large herbivores that are known to cause considerable damage to crop plants—wild pigs, deer and buffaloes.

A:ga beḷe beḷedu uṭṭu eraḍu varsha mu:ru varsha ma:ḍu uṭṭu, a:ne gine bandu nuggu uḍtu andale tindu uḍtittu adava.

Then (I) planted crops for 2 or 3 years, but if elephants and other animals came (to my field), they would eat everything.

As in many languages, the topmost (in the ethnoclassification hierarchy) superordinate categories of all plants and all animals tend to be unnamed, or 'covert', categories [9]. Nevertheless, Solega people do frequently make reference to these categories, and this is done by chaining together the chief members of each category, to form a loose compound.[1] Thus, the compound *mara maṇḍi giḍa* 'tree tree (a synonym) plant' or simply just *mara giḍa* 'tree plant' is often used to denote the category of all plants, while a similar construction, *hakki pera:ṇi* 'bird mammal' is used to refer to animals as a whole. The latter expression can be modified to refer to certain types of animals, such as in the following example, where the speaker is talking about the kinds of animals that live on the moss- and lichen-covered branches of the trees growing in shola (tropical montane) forest patches. The speaker uses the more elaborate expression *phalli pera:ṇi ha:vu sette-bodde* 'gecko mammal snake green-snake'. This is most likely a nonce coining, i.e. an expression that the speaker made up in response to a lexical need, whereas the shorter expressions *mara giḍa* or *hakki pera:ṇi* have been permanently incorporated into the Solega lexicon.

A: pa:seya me:le phalli pera:ṇi ha:vu sette-bodde ella baḷte ade
Geckos, snakes and other small animals all live on the moss.

3.7 Plants in Place Names

Studies on numerous languages have shown that plant names are frequently and productively incorporated into toponyms (place names) in for example, England [159, 160] the Middle East [161, 162], Hungary [163], and northern and southern Australia [164, 165]. The plants mentioned in Solega toponyms are also mostly trees, although the names of vines, shrubs and other prominent species, such as bamboo, are also sometimes incorporated. The following account of the relationship between Solega place names, plant names and named trees is an extract from a larger paper on Solega toponyms, published in the journal *Anthropological Linguistics* [166].

Place names in Solega typically have a bipartite X–Y structure, where Y is almost always a landscape or hydrological feature (see Table 5.2), although man-made features such as roads, reservoir embankments or dwellings can also occupy this slot. Examples include:

bu:di bare	'ash boulder'
kukkatti ku:ḍu	'name.of.vine crossroads'
basavana kaḍavu	'GOD.NAME's stream'

Position X is, in many cases, occupied by a plant name, as in the second example above. When X is the name of a god, or refers to a human entity (as in *basava-na kaḍavu* above, or *kaḷḷa-ra beṭṭa* 'thieves' mountain'), the GENitive marker *-na*, *-da*, *-ra* or *-ya* is affixed to X, to indicate that the head noun is associated with X. Rarely, a tree or animal name also takes the genitive marker, as in *te:gi-na kaṭṭe* 'teak embankment' or *gumma-na guḍḍe* 'owl hill'. When the head (the final element) of

[1] There are definite prosodic (intonational) and formal linguistic differences between such loose compounds and 'true' compounds such as *biḷi sampage mara* 'white *sampage* tree'.

the noun phrase is not a landscape feature, the referent of the toponym can be both a real-world object, as well as a place, as in *ne:ri haṭṭi* '*Syzygium cumini* shed'. This is also particularly true of the names of individual trees (see 'The naming of individual trees' below), such as *huli ha:rida honne* 'tiger jumped *Pterocarpus.marsupium*'.

Roughly a fifth of place names is tripartite (X–Y–Z). In some such cases, Y and Z are both landscape terms, while in others, the Y position is occupied by a verb, and the toponym comes to signify an action or event.

te:k-arre abbi	'teak rock.platform waterfall'
katte satta pa:la	'donkey died ford'
guṇḍu sikkida habbi	'bullet found waterfall'

3.7.1 Sources of Current Plant Distributions

Around 230 Solega place names have been recorded from the B. R. Hills, and of these, roughly one-third contains the name of a plant (Table 3.2). As described in [166], Solega toponyms that contain a plant name provide a record of an individual large tree or single-species group of trees. That these trees really do exist in such named locations can be readily verified, except in the case of trees that have died (see the following section for more details). Such traditional knowledge of the locations of individual tree specimens complements the Solega's otherwise very precise knowledge of the distributions of individual tree species. Practically all adult Solega can very accurately identify the forest types where any given tree species may be found.

The Solega language has numerous terms for forest or vegetation types that are distinguished by various features, such as altitude, topography, species composition and geology. These are described in more detail in Chap. 5. On being asked to describe a particular forest type, a Solega person will often provide a detailed list of the trees that typically occur in such a habitat. Such lists can provide a rough but accurate idea of not only the species composition of a particular location, but also the small-scale geographic distribution of a particular plant species. In the following extract, the speaker lists the tree species occurring inside the *ka:nu ka:ḍu* 'evergreen forest' (many of these trees have yet to be identified scientifically):

(BG) *Ka:ninalli ha:le mara ade, thuruve mara ade, kakkilu, bikkilu, bella:ḍe, ku:ma:ũ, kende, soravilu, aravilu, hebbe:vu... aravilu kende maradalli je:na:de.*
(In the *ka:nu*, there's the *ha:le* tree, the *thuruve* tree, etc.…there are bees/honey on the *aravilu* and *kende* trees.)

(MRM) *Nalavattu aivattu je:nu baratte onde maradalli.*
(Forty or fifty colonies come to a single tree.)

(BG) *Ne:ri, ne:ri mara,a:ga matte, i: ta:ri mara ondu ille. A ka:nu oḷage e:va ja:ga:dalli ue ta:ri mara huṭṭa:dille. A:mele, koḷamada mara, mi:nada mara...*

Table 3.2 Solega toponyms referring to plants

Solega toponym	English gloss
kallu ki:re haḷḷa	edible.green + stream
soreka:yi beṭṭa	cucurbit + mountain
nelli kadiru	Indian.gooseberry + grove
icci beṭṭa	fig + mountain
kaṭṭuguḷi gadde	*Ardisia* + marsh
meṇasu kaḍavu	pepper + stream
kumbaḷaka:yi beṭṭa	pumpkin + mountain
jo:ḷada bare beṭṭa	corn + boulder + mountain
ba:ḷe arre	banana + rock.platform
ja:ji arre	wild jasmine + rock.platform
gi:jigana tunḍa	*Caesalpinia mimosoides* + stump
se:bina ko:be	*Colocasia* + valley
se:bina kere	*Colocasia* + reservoir
gaṇigilu arre	*Plumeria* + rock.platform
battada gadde	paddy + marsh
kukkatti ku:ḍu	large.vine + crossroads
kaḍle bare	peanut + boulder
ji:rige gadde	wild.cumin + marsh
guḷḷada bailu	plant.name + clearing
ma:guḷi beṭṭa	plant.name + mountain
hanile arre	*Terminalia chebula* + rock platform
honne-mara bo:re	*Pterocarpus marsupium* + slope
honne arre	*P. marsupium* + rock.platform
honne bare	*P. marsupium* + cliff
honne gunḍi	*P. marsupium* + water.hole
kanchuva:ḷa dimba	*Bauhinia purpurea* + plateau
kithuḷi kaḍavu	orange.tree + stream
si:rimbi aḷḷa	thorny.lime + stream
ne:ri-gere	*Syzygium cumini* + reservoir
ne:ri ka:nu	*S. cumini* + evergreen.forest
ne:ri haṭṭi	*S. cumini* + shed
te:gina kaṭṭe	teak + embankment
muttagada gadde	*Butea monosperma* + marsh
si:ge va:ḍi beṭṭa	*Acacia sinuata* + tangle + mountain
si:ge guḷi	*A. sinuata* + pit
si:ge beṭṭa	*A. sinuata* + mountain
daḍasu mene habbi	*Grewia tilifolia* + house + waterfall
tenginamarada ko:bu	coconut.palm + slope
muttagada beṭṭa	*Butea monosperma* + mountain
atti haṭṭi	*Ficus racemosa* + shed
ko:li kaḍavu	*Wendlandia thyrsoidea* + stream
benḍe gunḍi	*Kydia calycina* + water.hole

(continued)

Table 3.2 (continued)

Solega toponym	English gloss
nelli beṭṭa	Indian.gooseberry + mountain
siguri haḷḷa	*Memecylon umbellatum* + stream
ja:la ga:ḍi bo:re	*Shorea roxburghii* + monoculture + slope
te:k-arre maṭṭa	teak + rock.platform + flat.land
te:k-arre abbi	teak + waterfall
thonḍ-egere	*Viburnum punctatum* + reservoir
doḷḷi kaṇuve	*Careya arborea* + pass
munḍu-halasu kaḍavu	stump + jackfruit + stream
ge:ru arre	cashew.tree + rock.platform
ko:li haṇṇi	*Wendlandia thyrsoidea* + pond
matti ni:ru gunḍi	*Terminalia arjuna* + water + hole
sa:gaḍe guḍḍe	*Schleichera oleosa* + hill
bejja arre	*Anogeissus latifolia* + rock.platform
seppe kaḍavu	*Entada rheedii* + stream
doḍḍa halasu gadde	big + jackfruit + marsh
erakina gadde	*Macaranga* sp. + marsh
kaggali guṇḍi	*Acacia chundra* + water.hole
kavaḷi kaṭṭe	plant.name + embankment
a:le haḷḷa	*Ficus* sp. + stream
maddaḷe ma:vu boḷi	drum + mango.tree + hilltop
sõgu ma:vu gadde	fragrant + mango.tree + marsh
ko:li arre beṭṭa	*Wendlandia thyrsoidea* + rock.platform + mountain
kaṭṭe ba:vi	senescent bamboo + well

(The *ne:ri* tree, and then, the *ta:ri* tree is the only one missing. There isn't a single *ta:ri* tree growing anywhere inside the *ka:nu ka:ḍu*. And then, there's the *koḷama* tree, the *mi:na* tree…)

Lists such as this provide valuable information on the kinds of species usually associated with a particular forest type, but toponyms that incorporate plant names allow a very precise determination of the existence of a plant species at a particular location. In many cases, the plants referred to in the toponym have some utilitarian significance, as in the case of *muṇḍuhalasu kaḍavu* 'stumpy.jackfruit stream', and *ja:ji arre* '*ja:ji*.jasmine rock.platform'. These denote places where one might find the important edible fruit and fragrant flower respectively.

3.7.2 Sources of Historical Biodiversity

Large, culturally important forest trees are often lost to storms, forest fires, insect damage or disease, but the place names that refer to their original location continue to be used. Solega communities are thus able to retain memories of these significant trees through toponyms, and preserve, in folk memory, what is effectively a historical record of large forest trees or even smaller plants (including crop plants) that are no longer present. One such example is *tenginamarada ko:bu* 'coconut tree valley', which recalls the fact that there used to be a British-initiated coconut plantation at this site, whose trees were later cut down for timber.

Other historical botanical events may also be encoded in place names, prominent among these being the flowering of plietesial species such as bamboo. These plants take many years to mature, flower only once in their lifetimes, and die shortly afterwards, having produced an abundance of seeds in a single burst. Clumps of dead bamboo are called *kaṭṭe* in Solega, and a small number of toponyms make reference to such clumps, as in *kaṭṭe ba:vi* 'senescent bamboo well'. The importance of such bio-historical information should not be underestimated, especially in the light of modern-day ecological problems facing the Solega's traditional lands. These include the rampant growth of the invasive woody weed *Lantana camara* over large swathes of the forest understorey, and the resultant suppression of local grasses, herbs and shrubs. Solega toponyms that commemorate dead trees may, in the not too distant future, provide valuable clues to the historical biodiversity of the forest, as they encode, with a high degree of precision, both the species and location of the named tree(s). This is especially relevant in the context of *Lantana* invasion, as most Solega are able to name many herbs and grasses that have now become locally extinct. For example, the toponym *kaũri ba:vi* refers to *Helicteres isora*, a common shrub (*kaũri* in Solega) that is said to be favored by elephants, and is now rare. Similary, the names of grasses like *sabakana hullu* and *nose hullu*, which were abundant when the Solega still practised their leaf litter fires, will in future probably exist only in Solega toponyms like *sabakana gadde* '*sabaka*.grass marsh' and *nose arre* '*nose*.grass rock.platform'.

Solega place names often make reference to crop plants which were once grown by the inhabitants of now-abandoned settlements. Examples include *soreka:ye beṭṭa* 'gourd mountain', *battada gadde* 'paddy marsh', *jo:ḷada bare beṭṭa* 'corn boulder mountain' and *ba:ḷe arre* 'banana rock.platform'. Others like *ne:ri haṭṭi* and *atti haṭṭi* (*haṭṭi*=cowshed) refer to historical practices when cattle in large numbers (belonging to non-Solega peoples) were brought up from the plains to the hills in summer when grass was scarce in the lower elevations. Even though these crops or structures are no longer to be found in these locations, the names indicate key periods in the bio-cultural history of these ancestral communities.

3.8 The Naming of Individual Trees

Individual trees are often singled out for recognition by the Solega, by virtue of their religious, utilitarian or perceptual properties. Like place names, tree names are also compound lexemes, but here the head noun (the final element) is the name of the tree species, while the modifiers can include nouns or adjectives that encode a salient attribute of the tree, or even a full toponym that marks the locality where the named tree is found. An example of the latter is *gumma-na guḍḍe ta:ri* 'owl-GEN hill *Terminalia.bellirica*', the name of a large *T. bellirica* growing on a specific hill which is, in turn, named after the owls that live on it. As is evident from Table 3.2, tree names can contain the names of other plant species, especially if the tree in question is associated with a plant-based toponym.

3.8.1 Sacred Trees

The most significant trees in the Solega religion are two large and very old individuals of the species *Michelia champaca*, which are known by the names *doḍḍa sampage* 'big *sampage*' and *cikka sampage* 'small *sampage*' (*sampage* is the Solega name for the *M. champaca* tree). These two trees are venerated as sacred entities in their own right, and are also associated with a myth involving the Hindu god and goddess Shiva and Parvati, who are supposed to have rested under the *doḍḍa sampage* tree. It is also claimed that a very large *huṭṭu linga* 'in-situ *lingam*' (a stone icon of Shiva) is to be found under the roots of this tree.

3.8.2 Bee Trees

The giant honeybee (*Apis dorsata;* Sol. *hejje:nu*) is an important resource to the Solega, both in a cultural and economic sense [148]. The honey of this and other bee species was traditionally consumed in large quantities by individual Solega while out foraging, travelling, or working their fields. Some of these bee species, and in particular *hejje:nu*, are migratory, and only arrive in the Solega's lands at the onset of the pre-monsoon rains, a migration that coincides with the flowering of large rainforest trees. According to the Solega, *hejje:nu* prefer to live in evergreen forests on very tall trees, showing a marked preference for the *soravilu* (*Acrocarpus fraxinifolius*) and *ba:ge/sele* (*Albizzia odoratissima, Albizzia lebbek*) trees. Individual trees of these species become known across a range of Solega settlements as *je:nu mara* or 'bee/honey trees', due to the fact that they are home to a large number of *hejje:nu* colonies year after year. For instance, the *do:vu ma:vu ba:ge* is a single large *Albizzia odoratissima* found near *ko:li ba:vi* hill, which attracts upto 50 hejje:nu colonies around the same time every year (Fig. 3.2). An invitation to go to *do:vu ma:vu ba:ge* at the right time of year can only mean a honey-harvesting trip; no further explanation is required. Bee trees are often named after the location in which they are found growing, although the physical features of the tree may also sometimes feature in the name (Table 3.3). Several other 'bee trees' are known to the Solega, and these, along with other locations where bees often nest, are remembered as a mental map of honey-harvesting sites.

Fig. 3.2 The bee tree known as *do:vu ma:vu ba:ge*. This photo was taken early in the wet season (April 12, 2012), when only five *Apis dorsata* hives had taken up residence

3.8.3 Trees Named for Their Physical Appearance

A number of trees are named for their unique physical characteristics, especially when these characteristics dictate the nature of Solega people's interactions with the trees. There may be considerable overlap between this category of names trees and the 'bee trees', as many members of the former category may also serve as homes for numerous *hejje:nu* colonies. Hence, *juṇṭu a:dire* and *seḷeyana honne*, whose names make reference to their shape and great height, encode the fact that it is very difficult to climb to the tops of these two trees (Fig. 3.3). This is particularly salient in the monsoon months, when the honeybee colonies nesting in the upper reaches of these trees may have to be left untouched by Solega honey gatherers. The name *koma:ra matti* arises from the presence of a door-like opening into the trunk of this large tree, while *uiya:le ta:re* derives its name from the fact that the two similarly sized *Terminalia bellirica* trees growing adjacent to one another serve as two posts to which an *uiya:le* 'swing' can be attached (Fig. 3.4a). The Solega believe that a *daiya*, a malevolent female spirit, uses these trees to swing from.

Table 3.3 Individual trees with Solega proper names

Religious	*dodda sampage*	big + *Michelia champaca*
	cikka sampage	small + *M. champaca*
Bee trees	*jo:du turuve*	twin + *Ligustrum perottettii*
	do:vu ma:vu ba:ge	? + mango + *Albizzia odoratissima*
	basappana kadavina aranelli	PERSON + stream + tree.name
	dodda kollada aranelli	big + pool + tree.name
	ittu bu:di soravilu	flour + ash + *Acrocarpus fraxinifolius*
	*si:ge guli ba:ge**	*Acacia sinuata* + pit + *A. odoratissima*
	*modeya:na ta:ri**	? + *Terminalia bellirica*
	*dodda sampage soravilu**	big + *Michelia champaca* + *A. fraxinifolius*
	*siguri halla soravilu**	*Memecylon umbellatum* + stream + *A. fraxinifolius*
	*gummana gudde ta:ri**	owl + hill + *T. bellirica*
	*du:pa mara** in *kambaligadde* evergreen forest	?
	*kende mara** in *kambaligadde* evergreen forest	?
Physical	*uiya:le ta:ri*	swing + *T. bellirica*
	juntu a:dire	tree.top + tree.name
	koma:ra matti	doorframe + *T. crenulata*
	seleyana honne	very.tall + *Pterocarpus marsupium*
Other	*midi ma:vu*	small.unripe.fruit + mango.tree
	tangulu ma:vu	leftovers + mango.tree
	huli ha:rida honne	tiger + jump + *P. marsupium*

Names followed by asterisks indicate trees that are now dead

3.8.4 Other Named Trees

Some trees have names even though they are not sacred or sources of honey, nor do they have any unique physical characteristics. *midi ma:vu* probably refers to a mango tree that has small fruit that are slow to ripen. The name *tangulu ma:vu*, on the other, gives an indication of the tree's origin. It is said that this mango tree sprouted from seeds discarded by people who had once rested here, and eaten a meal (*tangulu* is the Solega word for 'leftovers'). Finally, the tree named *huli ha:rida honne* 'tiger jumped *P.marsupium*' may have been named after regular sightings of the animal near it (see the following section for a more detailed discussion on toponyms that refer to animals). This is a tree with numerous claw marks on its bark, both old and new, and it is possible that a tiger might even rest on it, given its low, almost horizontal branches (Fig. 3.4b).

Fig. 3.3 The tree named *juṇṭu a:dire*, after the appearance of the flexible apical branches of this tree (*'juṇṭu'* is Solega for a tuft of hair). Photos by S. Agnihotri. Originally published in [166]

3.8.5 The Referent of a Named Tree

The names of individual trees make reference not only to the tree itself, but also to the general environs of the tree [166]. The most famous of these, *doḍḍa sampage*, is often referred to by Solega people as a landmark when describing other locations nearby, or when giving directions. This tree has also lent its name to a stream that flows past it (*doḍḍa sampage aḷḷa*), the patch of forest in which the tree is situated as well as to the nearby village which is home to the primary caretakers of the tree (*doḍḍa sampage po:ḍu*). While this village has an alternative, unrelated name (*keredimba po:ḍu*), it is sometimes referred to by its inhabitants simply as *doḍḍa sampage*, as this name can refer not only to the tree, but also to the land around it.

3.9 Plant Uses

"Ka:ḍa oḷagaḍe beka:dashṭu vidya ade," "There is much wisdom in the forest", said Heddini Basavegowda in one elicitation session, before proceeding to describe some of the natural remedies known to him. Much of this wisdom is shared by the general population, and practically every adult Solega person will be able to recite

a

b

Fig. 3.4 (**a**) The pair of *T. bellirica* trees named *uiya:le ta:re*. (**b**) The *P. marsupium* tree named *huli ha:rida* honne. The *white arrow* indicates a branch that a tiger might rest on. Photos by S. Agnihotri. Originally published in [166]

the medicinal (and other) uses of a number of forest plants. These, however, are what Hunn [147] would characterise as 'simple remedies', whereby the appropriate part of a single plant is consumed in the form of a suitable preparation to treat a particular ailment. 'Complex remedies' do exist, but are known in large numbers only to specialist practitioners. While there does not appear to be a Solega word that designates someone as a 'medicine man/woman' the knowledge of complex remedies tends to belong to senior community members, such as priests and village council members. Table 3.4 below lists the many uses that the plants of the B. R. Hills are put to by the Solega—they include consuming plants as food or medicine or using them to make artefacts, just to name a few. The remedies listed under 'medicine' are all simple remedies that Solega people will readily share with outsiders or with one another. While several complex remedies have been documented by the author, these have been omitted from Table 3.4 as the knowledge owners have deemed them unsuitable for the public domain.

Plants utilised as food form the largest category in Table 3.4, closely followed by medicinal plants. One of the more interesting aspects of Table 3.4 is the presence of numerous reported uses of plants by non-human species, including birds, honeybees and a range of large and small mammals. Of the plants mentioned above, those that Solega people have noted as being used by other animals (18 %) rival in number the plants used by humans as medicines (21 %) and as food (22 %). This category also has implications for the 'intellectual' vs. 'utilitarian' debate addressed by publications such as [62] and [9] (see below). A further 5 % of plants serve as indicators of important phenological or meteorological phenomena.

The observation that so many plant–animal interactions are noticed and commented upon, can be explained in two ways. One, that knowing the food habits of birds, bees and wild mammals may indirectly be of use to humans. The foraging habits of honeybees constitute the most straightforward aspect of this phenomenon—the presence of large numbers of bees on a particular type of flower signals not only the imminence of a honey-flow, but also the type of honey that will soon be available for harvesting, as well as providing clues to the location of the bees' hive(s). This is made explicit in the following passage:

> *Hu: kaḑame ira dina, je:nu sariya:gi kelsa ma:ḑalla. Honne hu: ja:sti a:da dina hu:vina me:le ja:sti kelsa ma:ḍutte. A maradalli je:nu iddare, ishṭu du:radalli no:ḑkobahudu, e:kandare hu:vinda manege ja:sti o:ḑa:ḑtare... Namma avaru e:na ma:ḑtare, honne hu: banda dina ka:ḑiga oitare. E:kandare je:nu sigutte a:ga. Honne hu: ho:da:ga maneli kelsa ma:ḑta:re.*
>
> On days when there are few flowers, the bees don't work properly. When the *honne* flower is fully in bloom, the bees work a lot. If there are bees on a tree over there [POINTS], you can see them from here, because large numbers fly between the flowers and the hive... What our people do is, they go to the forest when the *honne* flower is in bloom. Because that's when you get honey. When the *honne* flowers go away, people work at home.

Second, Solega men used to hunt various birds and mammals for food and medicinal products (hunting no longer occurs within the confines of the BRT wildlife sanctuary), and an important part of the skill set of any hunter would have been a precise knowledge of the behaviour, distribution and feeding habits of prey species.

Table 3.4 Some plants and their relationship with Solega people

Solega name (complete)	Scientific name	Forest type	Use
Food			
arre aṇṇe	*Bryophyllum* spp., *Kalanchoe* spp. *Commelina ensifolia (?)*, *Cyanotis* sp., *Habenaria longicornu*		Species with yellow leaves is eaten
beṇṇe mara	*Persea americana*		Exotic tree; fruit can be eaten
beṭṭa maruḷi giḍa	*Indigofera wightii*		The honey made from this flower's nectar is highly sought after
bikkilu mara#	*Elaeocarpus serratus*	*male, ka:nu*	Seeds are eaten
biḷi ju:jakki	*Lantana indica*	*na:ḍu*	Fruit can be eaten
biḷi ko:muḷḷi giḍa	*Rubus niveus*	*male*, sometimes *na:ḍu*	Edible fruit ripen in the early monsoon
bu:ra da mara#	*Bombax ceiba*	*na:ḍu*	Flower buds are eaten
cokkattãna mara, cokkottã	*Citrus* sp.	*male*	Edible bittersweet fruit ripen in the monsoon
daḍasina mara#	*Grewia tiliifolia*	*male, na:ḍu* (smaller variety)	Small edible fruit ripen in July
doḍḍa puḷḷuḷi giḍa	*Oxalis* sp.	*male*	Leaves and stems have a sour taste, and are cooked in *uduka* 'soup'
gi·jiga·na ambu	*Caesalpinia mimosoides*		The young shoots are eaten
goṭṭi ambu#	*Ziziphus rugosa/ xylopyrus?*	*male*, sometimes *na:ḍu*	Edible fruits ripen in April–May
honne mara#	*Pterocarpus marsupium*	*male*	Honey from this tree is very strong tasting
iṭṭu nelli mara#	*Emblica indofischeri*	*na:ḍu*	Fruit are eaten raw or pickled
ju:jakki	*Scuttelaria* sp.		Ripe fruit are eaten
ka:ḍa ge:ru mara	*Semicarpus anacardium*	*male, na:ḍu*	Entire ripe fruit is eaten (July–August)
ka:ḍa se:bu giḍa	*Colocasia* sp.		Edible corms
ka:ḍasuṇṭi giḍa#	*Zingiber officinale*		Wild ginger, cooked with food for flavouring
ka:ge ma:ri ambu	*Passiflora edulis*	*male, ka:nu*	Edible fruit ripen in August
kakkilu mara	*Celtis tetrandra*	*male, ka:nu*	Edible fruit ripen in September
kancuva:ḷa mara#	*Bauhinia purpurea*	*male, na:ḍu*	Seeds are eaten, the gaṇḍu (male) plants produce smaller, inedible fruit
karava:di mara#	*Persea macrantha*	*male, ka:nu*	Young leaves are cooked in *uduka*

(continued)

Table 3.4 (continued)

Solega name (complete)	Scientific name	Forest type	Use
kaṭṭaguḷi giḍa	*Ardisia solanacea*	*male, ka:nu*	Stems are chewed for a sour taste
komba:le mara#	*Wrightia tomentosa*	*male, na:ḍu*	Flowers (June–July); added to milk to make curd
lakoṭi mara	*Polygala arillata*	*male, na:ḍu*	Exotic plant that now grows in forests; fruit (monsoons) are eaten by people
matti mara#	*Terminalia crenulata*, *T. elliptica* (syn.), *T. tomentosa* (syn.)	*male, na:ḍu* infrequently in *ka:nu*	The flowers yield viscous honey
ma:vũ, torre ma:vũ/mara#	*Mangifera indica*	*male, ka:nu* also *na:ḍu* near streams	Edible fruits
na:yĩ bela/sujjilu mara	*Acacia leucophloea*		Alcohol made from the bark
nagaṛe mara	*Commiphora* sp., *Balsamodendron* sp.	*na:ḍu*	Edible fruits appear in the monsoon (June)
na:yĩ nelli mara	*Phyllanthus emblica*		Fruit are eaten raw or pickled
na:yĩbela mara	*Naringi crenulata*	*na:ḍu*	The bark is used by plains people to brew alcohol
nela go:ṇi soppu	*Bacopa monnieri*	*na:ḍu*	Plant is eaten
nela javanada ambu	*Cyanotis arachnoidea*	*na:ḍu*	Leaves are eaten
neve geṇasu	*Dioscorea glabra*		Yam whose long tuber grows vertically down; the superior tip of the tuber is fibrous and inedible
pa:pasagaḷḷi	*Opuntia* sp.	*na:ḍu*	Edible fruit ripen in August
sa:gaḍe mara#	*Schleichera oleosa*	all types	Fruit (July) are edible
seḷḷe mara#	*Cordia obliqua*	*male, na:ḍu*	Edible fruit ripen in the monsoon (June–July)
thu:ba:re mara#	*Diospyros melanoxylon (?)*	*na:ḍu*	Edible fruits ripen in July
to:ṭambu	*Dioscorea* sp.		Yams are ready to eat in January
u:li, uḷinelli giḍa#	*Phyllanthus reticulatus*, *Kirganelia reticulata* (syn.)	*male, na:ḍu*	Small fruit can be pickled
uṇise mara	*Tamarindus indica*		Used to impart a sour flavour to cooked food
uppilu	*Taxillus tomentosus*		Big mistletoe that grows on daḍasu trees are eaten
belare geṇasu	*Dioscorea oppositifolia*		Edible yam

(continued)

Table 3.4 (continued)

Solega name (complete)	Scientific name	Forest type	Use
beppa:le mara#	*Wrightia antidysenterica*		Latex added to milk to make curd
doḍḍa ka:re mara	*Canthium dicoccum*	*male, ka:nu*	Edible fruit
gaṇake#	*Solanum nigrum*		Edible fruit
gumuṭi giḍa#	*Nicandra physaloides*		Edible fruit of the big (doḍḍa) form ripen in the dry season, and those of the small (saṇṇa) form ripen in the monsoon
ku:gu giḍa	*Schumanianthus virgatus*		Leaves for baking roṭṭi
ne:raḷe mara	*Syzygium cumini*		Edible fruit
handi neve geṇasu	*Dioscorea bulbifera*		Edible yam
nu:re geṇasu	*Dioscorea hispida*		Edible yam
e:ḷakki giḍa	*Elettaria cardamomum*		Cardamom, used in cooking
Indicator			
beṇḍe mara#	*Kydia calycina*	*male*	Flowering (September) indicates a time of the year when langurs have large fat reserves; the honey from this plant is yellow and tastes good
bikkilu mara#	*Elaeocarpus serratus*	*male, ka:nu*	Appearance of the fruit (July–August) indicates the time of the year when deer (*kurri*) and mouse deer have large fat reserves
iṇḍãna giḍa	*Pterolobium hexapetalum*	*na:ḍu*	Commonly home to the dwarf bee
kaũri mara#	*Helicteres isora*	*male, na:ḍu*	Flowering of this tree indicates a time of the year when langurs have large fat reserves
kilanji mara	*Chukrassia tabularis*	*male, ka:nu*	Frequently a home to the giant honeybee
ku:ḷi mara#	*Gmelina arborea (asiatica?)*	*male,* sometimes *na:ḍu*	Appearance of fruit (February) indicates time of the year when these animals have large fat reserves
na:yĩ gaṇṭu mara	*Aphanamixis polystachya*	*ka:nu*	Flowers in the early monsoon; usually home to bees
sele mara	*Albizia odoratissima*	all types	Flowers in the early monsoon; coincides with the appearance of wax moths in beehives (as well as spoilt honey)

(continued)

Table 3.4 (continued)

Solega name (complete)	Scientific name	Forest type	Use
sipure giḍa #	*Asparagus* sp.	all types	Appearance of the fruit indicates time of year when monitor lizards have large fat reserves
thuruve mara	*Ligustrum perrottetii*	*ka:nu, shola*	Home to *hejje:nu* and *thuruve je:nu bees*
u:li, uḷinelli giḍa#	*Phyllanthus reticulatus, Kirganelia reticulata* (syn.)	*male, na:ḍu*	Commonly home to the dwarf honeybee
mustakada mara	*Meliosma pinnata (?)*		First flower of the monsoons, and a sign of the approaching rain
Medicine			
araḷi mara	*Ficus religiosa*	*na:ḍu*	Roṭṭi is baked in these leaves and given to children who can't speak properly
araḷu giḍa	*Ricinus communis*	*male, na:ḍu*	Oil is used as hair-oil, rubbed into scalp for headache or into joints for arthritic pains
aranelli mara	*Garuga pinnata*	*male, na:ḍu* sometimes *ka:nu*	Fruit eaten by animals; bark is group up and applied to the scalp in case of localised headaches
atti mara	*Ficus racemosa*	*male, ka:nu,* sometimes *na:ḍu*	Edible fruit, good for eyesight
a:guri mara	*Trema orientalis*	all types, but infrequently in *na:ḍu*	Bark is ground up, and extract drunk in cases of diarrhoea
basale ambu	*Basella alba*	*na:ḍu*	In case of mouth ulcers or a sore throat, the leaves and stem are consumed and a string made from the fibres is worn around the neck
bejja da mara	*Anogeissus latifolia*	*male, na:ḍu*	Resin is eaten to increase appetite, leaves are often attacked by smooth white caterpillars
cennãgi giḍa#	*Gardenia turgida*	*na:ḍu*	Waxy covering of young leaves is melted in castor oil, and applied to the heads of children who have a cold
doḍḍa a:luppe mara	*Wrightia tinctoria*	*ka:nu*	Latex from crushed leaves is placed on abcesses
doḍḍa uttara:ṇe giḍa	*Stachytarpheta indica*	*male*	Paste of the root is applied to the body of children suffering from 'weakness' (eḷavu)

(continued)

Table 3.4 (continued)

Solega name (complete)	Scientific name	Forest type	Use
gandada mara	*Santalum album*		Paste is applied to the side of the face when swollen during toothache
gore uṇṇina giḍa, goruṇṇina giḍa	*Plectranthus* sp.	*male, na:ḍu*	Juice from the leaves is applied to the webs of fingers and toes to cure tinea
goṭṭi ambu#	*Ziziphus rugosa/ xylopyrus?*	*male,* sometimes *na:ḍu*	Bark is mixed with the root of handi ba:ṇe hullu and drunk in case of stomach ache
honne mara#	*Pterocarpus marsupium*	*male*	The sap is used to help broken bones heal; applied to the inside of a bamboo splint
hoṭṭeno:vina ambu	*Cissampelos pareira*		A remedy for stomach ache
iṭṭu nelli mara#	*Emblica indofischeri*	*na:ḍu*	Bark extract is applied to skin that has blistered after contact with the spines of ele se:ḷu caterpillar
ka:ḍa arasina giḍa	*Curcuma longa*		Paste of the root is applied to the face to treat pimples
kaḍabada mara	*Mitragyna parviflora*	*male, na:ḍu*	Bark extract is given to dogs prior to hunting to improve alertness and tracking ability
ka:ḍasuṇṭi giḍa#	*Zingiber officinale*		Added to tea as a cure for a common cold
kaĩye suṇḍe giḍa	*Solanum violaceum*	*male, na:ḍu*	Fruit (avail. Year-round) is used to treat common cold in young babies (6 months–1 year). Black peppercorns are ground to a fine powder and placed on the fontanelle of the head, and the juice of the kaĩye suṇḍde fruit is squeezed onto the same spot
kaĩyebe:vu mara, aḷavilu	*Azidirachta indica*	*na:ḍu*	Leaves are eaten in case of stomach ache or as a vermifuge
kakke mara#	*Cassia fistula*	*male, na:ḍu*	Bark extract is given to people suffering from gall and kidney stones
karava:di mara#	*Persea macrantha*	*male, ka:nu*	Bark extract is given to sick goats
kaũri mara#	*Helicteres isora*	*male, na:ḍu*	Pieces of the unripe fruit are tied to the wrists of restless infants whom are unable to sleep—the hand depends on the side that the infant usually faces while lying down

(continued)

Table 3.4 (continued)

Solega name (complete)	Scientific name	Forest type	Use
kende mara	*Elaeocarpus tuberculatus*	*male, ka:nu*	Bark is mixed with the root of iḷisingi vine, and the paste smeared on the body or head in case of fever, headache or fatigue
kittarsã mara	*Bauhinia racemosa*	*na:ḍu*	Bark of the tree is used as a contraceptive, the extract being ingested by both men and women.
kuḍuma:vũ/ ku:ma:vu/ kuḍuma:vu mara	*Cinnamomum malabatrum*	*ka:nu*	Powdered bark is mixed with *ra:gi* flour and eaten in case of diarrhoea; also a good remedy for fatigue
ku:ginele giḍa	*Canna* sp.	*ka:nu, shola*	Leaves are used to bake roṭṭi; said to help children develop the power of speech
meṇasi:ge mara	*Litsea josephii*	*male, ka:nu*	Seeds (powdered) and bark (chewed) are taken in case of appetite loss
na:yĩ nelli mara	*Phyllanthus emblica*		Bark extract applied to skin that has blistered after contact with the spines of ele se:ḷu
nela nelli giḍa	*Phyllanthus niruri*	*na:ḍu*	Leaves are eaten, esp. as a remedy for jaundice
nelama:gaḷi ambu	*Hemidesmus indicus*	*male, na:ḍu*	Root is consumed as a remedy for stomach ache
omma:lika giḍa	*Spilanthes* sp.	*male*	Leaves are crushed, and placed in the ear in case of earache; mixed with honey and consumed in case of a sore throat
onge mara	*Pongamia pinnata*	*na:ḍu*	Oil from seeds or bark extract is applied to the skin in case of a rash or blister
onnegã giḍa	*Lindernia antipoda*	*male, na:ḍu*	Leaves are added to uduka and eaten; leaves are consumed in case of jaundice
ro:jiga giḍa#	*Lantana camara*	*male, na:ḍu*	Crushed leaves are placed on wounds; the charcoal from woody branches is used as toothpaste
se:bina geṇasu#	*Colocasia* sp.		Stalks are eaten in uduka, useful as a vermifuge
seppe ambu#	*Entada rheedii*	*male, ka:nu*	The white seeds are ground up and the paste applied to abcesses

(continued)

Table 3.4 (continued)

Solega name (complete)	Scientific name	Forest type	Use
sipure giḍa#	*Asparagus* sp.	all types	Root is eaten as remedy for beestings; juice is drunk as a vermifuge
ta:re mara#	*Terminalia bellirica*	all types	Unripe fruit of this tree is mixed with unripe fruit of the castor plant and honey; taken as a remedy for cough
udurambe ambu	*Clematis gouriana*	*ka:nu*	In case of high fever or *u:dalu* 'swelling', the leaves are crushed and applied to the skin, all over the body
urigilu mara#	*Chloroxylon swietenia*	*na:ḍu*	Leaves are used on skin infections and rashes
a:le ambu	*Tylophora indica*		The latex is applied to scars, in order to darken discoloured skin
ambaṭe mara	*Spondias pinnata*		Young leaves are mixed with garlic, pepper and chilli, and the decoction given to women at the time of delivery
araḷe mara	*Terminalia chebula*		The unripe fruits are ground up, and smeared on head of a sick child
gaṇake giḍa#	*Solanum nigrum*		Juice of the leaves is a remedy for cough
ganṭe mara	*Schrebera swietenioides*		The dry fruits are ground up and given to children for cough
suḷḷi giḍa	*Curcuma* sp.		The juice of the underground stem is applied to wasp stings
maḍḍi mara	*Morinda tinctoria*		The resin is used as a remedy for cold in children
Used by other species			
a:la mara	*Ficus benghalensis*	*male, na:ḍu*	Fruits eaten by birds
ambe mara	*Canthium travancoricum*	*ka:nu, male,* sometimes *na:ḍu*	Fruits eaten by birds
benḍiga:na ambu	*Embelia tsjarium–cottam*	all types	Large inflorescences are visited by bees
bikkilu mara#	*Elaeocarpus serratus*	*male, ka:nu*	Visited in large numbers by bees
biḷi si:ge ambu	*Acacia sinuata*	*male, na:ḍu*	A preferred food of elephants
biḷicci mara	*Ficus amplissima*	*male, ka:nu*	Fruits ripen (June–July), eaten by birds

(continued)

Table 3.4 (continued)

Solega name (complete)	Scientific name	Forest type	Use
biṭe mara	*Dalbergia latifolia*	all types	One of the first trees to flower in the early monsoon, sticky, sweet exudate drips from the flowers onto the ground; visited in large numbers by *hejje:nu* honeybees
daḍasina mara#	*Grewia tiliifolia*	*male, na:ḍu* (smaller variety)	Elephants eat the bark, fruits are eaten by barbets
doḍḍa ugune ambu	*Ipomoea campanulata*	all types	Eaten by pigs and elephants
doḷḷi mara	*Careya arborea*	*male*	Fruits which ripen in the monsoons (June–July) are eaten by elephants; they are often fermented, and cause the elephants to appear drunk
gokkoruḷe ambu, gangonge ambu, kokkoro:ḷe/ gongoroḷe/huriana hambu/huriyã	*Diploclisia glaucescens*	*male*	Fruits eaten by bears
goṇḍe more mara	*Litsea floribunda*		Flowers are visited by bees and yield copious amounts of honey
go:ṇi mara	*Ficus drupacea*	all types	Fruits eaten by the kuggilu bird (dry season—April to May)
ka:ḍaseṇabbã giḍa#	*Crotolaria pallida*	*male*	The flowers (early monsoon) are visited by a smaller species of bees
kakke mara giḍa#	*Cassia fistula*	*male, na:ḍu*	Bark is eaten by porcupines; unripe fruit eaten by bears
kallicci mara	*Ficus parasitica*	*na:ḍu*	Fruits (June–July) are eaten by birds
ka:na gojjali mara	*Trichilia connaroides*	all types	Fruits eaten by birds
kappali mara	*Flacourtia indica*	all types	Fruits (July) eaten by birds
karava:di mara#	*Persea macrantha*	*male, ka:nu*	Fruits eaten in large quantities by nesting barbets
kari icci mara	*Ficus tsjahela*	*male, ka:nu*	Fruit ripen in the monsoons (June–July), and are eaten by birds
karigilu mara/ karagilu#	*Callicarpa tomentosa*	*male*	Flowers are visited by bees
ke:silu	*Mallotus phillippensis*	all types	Eaten by elephants
ku:ḷi mara#	*Gmelina arborea (asiatica?)*	*male,* sometimes *na:ḍu*	Fruits eaten by deer (*kurri*) and mouse deer

(continued)

Table 3.4 (continued)

Solega name (complete)	Scientific name	Forest type	Use
ku:re pandi giḍa	*Curculigo orchidoides*	all types	Bulbs are dug up and eaten by pigs
magga:re mara#	*Randia dumetorum*	*male, ka:nu*	Porcupines like to eat the bark
ma:vũ, torre ma:vũ/mara#	*Mangifera indica*	*male, ka:nu* also *na:ḍu* near streams	Fruits eaten by elephants
muḷḷu kappa:li#	*Scolopia crenata*	all types	Fruits (July) are eaten by birds
naravalu mara	*Olea glandulifera. O. paniculata* (syn.)	*male, ka:nu*	Flowers are visited by large numbers of bees
navila:ḍi mara	*Vitex regunda (?)*	*na:ḍu*	Flowers in the early monsoon, visited by smaller bee species
na:yĩbela mara	*Naringi crenulata*	*na:ḍu*	Unripe fruit can be poisonous to wild animals, such as deer
nere mara	*Vepris bilocularis*	*male, ka:nu*	Fruits (July) are eaten by birds; flowers are visited by bees
ni:ḷa:lu mara	*Bischofia javanica*	all types	Fruits (July/August) are eaten by birds; flowers are visited by bees; leaves are eaten by the giant and flying squirrels
sa:gaḍe mara#	*Schleichera oleosa*	all types	Fruits (July) are eaten by animals
taṇiki giḍa#	*Maesa indica*	*male, ka:nu*	Fruits (July) eaten by bears and birds
te:kina mara	*Tectona grandis*	*male*	Flowers (June/July) visited in large numbers by stingless bees; this is a good time to harvest stingless honey
toṭṭi mara	*Capparis cleghornii*	*na:ḍu*	Fruits eaten by bears
gumuṭi giḍa#	*Nicandra physaloides*		Fruits also eaten by snakes (especially the cobra)
icci mara	*Ficus nervosa*		Fruits eaten by birds
kincaga mara	*Erythrina indica*	all types	Birds and small animals drink rainwater out of the cup-shaped flowers
oluge mara	*Terminalia paniculata*	*male*	Flowers have a soporific effect on bees; said to give bees a headache, and make them stop flying
soravilu mara	*Acrocarpus fraxinifolius*		Bees like the flowers

(continued)

Table 3.4 (continued)

Solega name (complete)	Scientific name	Forest type	Use
kiribidiru	*Dendrocalamus strictus?*	*na:ḍu*	Large mammals eat the stems when they are sick
Artefacts			
anase mara	*Glochidion velutinum* (ka:na anase), *Glochidion zeylanicum*		Flowers in the monsoons (June–July); leaves are used to make smoking torches for honey gathering
bu:ra mara#	*Bombax ceiba*	*na:ḍu*	Silky fibres around the seeds are used as mattress and pillow stuffing
cennãgi giḍa#	*Gardenia turgida*	*na:ḍu*	Waxy covering of young leaves used to stick objects together (e.g. in *ta:li* necklaces)
je:nerakina mara	*Mallotus tetracoccus*	*male, ka:nu*	Leaves are used for baking roṭṭi, and as plates
kaĩyerakina mara	*Clerodendrum viscosum*	*male, ka:nu*	Leaves are used to make the receptacles giṇṇi and joṭṭe
kallu sauwe mara, i:rananna mara	*Frimiana colorata*	*male, na:ḍu*	Fibres from the trunk are used for making rope
karigilu mara/ karagilu#	*Callicarpa tomentosa*	*male*	Leaves used to make plates
ka:ta:ḷe mara	*Eriolena quinquelocularis*	*male*	Wood used to make handles for hoes and axes
komba:le mara#	*Wrightia tomentosa*	*male, na:ḍu*	Wood used to make a millet dough stirrer
se:bina geṇasu#	*Colocasia* sp.		Leaves are used as a receptacle for honey
seḷḷe mara#	*Cordia obliqua*	*male, na:ḍu*	Resin from the fruit is used as glue
urigilu mara#	*Chloroxylon swietenia*	*na:ḍu*	Wood used to make ploughs
beppa:le mara#	*Wrightia antidysenterica*		Wood used to make a millet dough stirrer
kallu muttaga giḍa	*Desmodium oojeinense, Ougeinia oojeinensis* (syn.)		Leaves used as containers
kallu sauwe mara	*Sterculia villosa*		Leaves used as containers
hebbidiru	*Bambusa arundinacea* or *Dendrocalamus strictus?*	*male*	Large stems are chopped into segments, and used as water carriers
Ceremony			
aḷḷa ko:li giḍa	*Ludwigia octovalvis*	*male*	Flowers used in festivals
biḷi sampage mara, keṇḍa sampage mara	*Michelia champaca*	*male, ka:nu*	Flowers used in religious ceremonies

(continued)

Table 3.4 (continued)

Solega name (complete)	Scientific name	Forest type	Use
gersi mallige giḍa	*Hypericum mysorense*	*na:ḍu*	Fragrant flowers used in religious ceremonies
ja:lada mara#	*Shorea roxburghii*	*male, na:ḍu*	Fragrant resin is used as incense; fragrant flowers (March–April) are used in religious ceremonies; wood is used to make doorframes
ka:na:du:pa mara	*Canarium strictum*	*ka:nu*	Solidified resin is used as incense
kancuva:ḷa mara#	*Bauhinia purpurea*	*male, na:ḍu*	Flowers used during religious ceremonies
kaṇṇi:ru mara	*Cassine paniculata*	*male, na:ḍu*	Bark is placed on a rock platform during a ceremony performed to stop rain
kappali mara	*Flacourtia indica*	all types	The bark and spines are used in religious ceremonies
kempu ekka	*Asclepias curassavia*	*male, ka:nu*	Flowers used in festivals
ke:silu	*Mallotus phillippensis*	all types	Colour from the unripe fruit was used in earlier times as *kumkuma* (to make a red mark on the forehead)
muḷḷu kappa:li#	*Scolopia crenata*	all types	The bark and spines are used in religious ceremonies
na:ḍa seluge du:pa	*Boswellia serrata*	*na:ḍu*	Fragrant flowers (dry season, April–May) are used in religious ceremonies
na:ga betta	*Calamus* spp.		Decorated stems of the three species of this plant are used in temples or at festivals
na:gasampage mara	*Couroupita guianensis*		An exotic tree; flowers used in religious ceremonies
ni:lilakki mara	*Vitex negundo*	*male, na:ḍu*	Leaves used in decorations in temples with tuḷasi leaves
panni:ru mallige#	*Portulaca grandiflora*	*male*	Flowers (monsoon) used in religious ceremonies
ta:re mara#	*Terminalia bellirica*	all types	Shrines to shani de:varu can only be situated at the base of this tree
ba:ge mara	*Albizzia amara*		Flowers offered at temples
beṭṭa ko:li mara	*Wendlandia thrysoidea*		Flowers offered at temples
biḷiekka giḍa	*Asclepias physocarpa*		The taproot is used for divination
du:pa mara	*Canarium strictum*		Resin used for incense
pa:paṭe mara	*Pavetta indica*		Flowers offered at temples

(continued)

Table 3.4 (continued)

Solega name (complete)	Scientific name	Forest type	Use
seluge du:pa, du:pa, na:ḍu du:pa mara	*Soymida febrifuga*		Incense
Construction			
biḷi ko:li giḍa	*Colebrookea oppositifolia*	*male, ka:nu*	Wood used for building fences and roofs, because the stems grow straight
goṇḍe hullu	*Themeda triandra*		Used in construction
ja:lada mara#	*Shorea roxburghii*	*male, na:ḍu*	Wood used to make doorframes
kaggali#	*Acacia chundra*	*na:ḍu*	Wood used to build fences
udi mara	*Radermachera xylocarpa*	*male*	Wood used for construction
thoṇḍe mara	*Viburnum punctatum*		Wood for construction of small huts
Firewood			
kaggali#	*Acacia chundra*	*na:ḍu*	Used as firewood
matti mara#	*Terminalia crenulata, T. elliptica* (syn.), *T. tomentosa* (syn.)	*male, na:ḍu* infrequently in *ka:nu*	Used for firewood during the monsoons, as it burns easily
taṇiki giḍa#	*Maesa indica*	*male, ka:nu*	Wood is slow-burning
ka:naraḷu mara/ nere	*Evodia lunu-ankenda*		Firewood
Noxious			
doḍḍa oṇṭe giḍa	*Cosmos bipinnatus*	all types	Considered a weed
doḍḍatursã giḍa	*Girardinia urticifolia*	*male, ka:nu*	Fruits and leaves are covered in irritating spines
ko:ḷi kuṭuma giḍa, kaṇṇu no:vina giḍa/daiyya mallige giḍa	*Gloriosa superba*	*male, na:ḍu*	Poisonous if the tuber is eaten
ro:jiga na giḍa#	*Lantana camara*	*male, na:ḍu*	Thorny bush that creates impenetrable thickets
unugana hullu	*Heteropogon contortus*	*na:ḍu*	Seeds can become embedded in the skin and cause infection
hattoṇṭe giḍa	*Cynoglossum furcatum*	*male, na:ḍu*	The burred fruits can pierce the skin and cause infections
Ornament			
bikkilu mara#	*Elaeocarpus serratus*	*male, ka:nu*	Seeds included in necklaces
gongoroḷe giḍa	*Polygonum* cf. *hydropiperoides*		4–5 Inflorescences of the female form are tied in a bundle with *ne:ri mara*, and placed above the door of the house (ornament)

(continued)

Table 3.4 (continued)

Solega name (complete)	Scientific name	Forest type	Use
ja:jimallige ambu	*Jasminum azoricum*	*male*	Fragrant flowers can be given to women to woo them
kurrugu giḍa	*Strobilanthes* spp.	*ka:nu*	Flowers once every 7 years
minciga giḍa	*Exacum tetragonum*		Fragrant flower used in the courting of women
panni:ru mallige giḍa#	*Portulaca grandiflora*	*male*	Flowers (monsoon) used as ornaments among older women
Other			
gaṇṭe giḍa	*Brugmansia candida*		Grown as a hedge plant
kaḷḷi giḍa	*Synadelium grantii*	*na:ḍu*	Grown as a hedge plant
koṭṭaraḷu giḍa	*Jatropa* sp.	*na:ḍu*	Grown as a hedge plant
thu:ba:re mara#	*Diospyros melanoxylon (?)*	*na:ḍu*	Leaves are used for making *bi:ḍi* cigarettes
bu:ta:ḷe mara	*Givotia rottleriformis*	*na:ḍu*	A piece of bark is tied onto small children to protect them from evil spirits
jagaḷa gaṇṭi mara	*Diospyros montana*	*male, na:ḍu*	Sorcery
pe: udi mara	*Oroxylum indicum*		Bark protects children from *de:vati* spirit
Soap			
aṇṭuva:ḷa da mara	*Sapindus laurifolius*	*male, ka:nu*	The unripe fruits are used to wash clothes
beṇḍe mara#	*Kydia calycina*	*male*	Small branches are crushed and used as soap to wash hair
daḍasina mara#	*Grewia tiliifolia*	*male, na:ḍu* (smaller variety)	Small branches are crushed and used as soap to wash hair
magga:re mara#	*Randia dumetorum*	*male, ka:nu*	Unripe fruit is used as soap for washing clothes
seppe ambu#	*Entada rheedii*	*male, ka:nu*	Bark is used as soap for washing clothes and hair
Toy			
ka:ḍaseṇabbã giḍa#	*Crotolaria pallida*	*male*	Children play with the dried fruits, which make a rattling sound
karaṇa kuṇḍala giḍa	*Impatiens* sp.	*male*	Children play with the dehiscent fruits and with the earring-shaped flowers

(continued)

Table 3.4 (continued)

Solega name (complete)	Scientific name	Forest type	Use
pa:riva:ḷada mara	*Butea monosperma*		Children play with the seeds; they are rubbed vigorously on a rock to heat them up before pressing them onto the victim's skin
poṭle giḍa	*Passiflora calcarata*		Children play with the fruit by popping them between their fingers

The plants listed above are those for which the author has been able to determine a reliable scientific identification *and* record a traditional use. Therefore, they represent a subset of all plant names documented in the Solega language. # indicates a plant that belongs to more than one use category. Forest types are described in detail in Chap. 5, but briefly, *ka:nu* can be translated as 'evergreen forest', *male* as 'highland deciduous or mixed forest' and *na:ḍu* as 'lowland forest/scrub'

This applies equally to animals that the Solega might wish to avoid because of the danger they pose to humans—knowing the preferred plant foods of bears, wild pigs, elephants and gaur would allow Solega people to be more alert when in the vicinity of these plants. There is, of course, a non-utilitarian aspect to this knowledge. I argue in Chap. 6, in the section 'Relationships' that Solega characterisations of the behaviour of animals vividly illustrate the great affinity that people feel with non-human organisms. The transcripts in Chap. 6 show that such characterisations frequently include comments on animals' feeding and self-medication habits. Solega people are keen observers of plant–animal interactions not only because such knowledge is frequently useful to humans, but also because the humans identify with the animals in question.

Chapter 4
Solega Ethno-ornithology

4.1 Introduction

The BRT forests are said to be home to approximately 250 species of birds [82, 83], and a significant number of bird species have been given names in the Solega language. This chapter presents collaborative research carried out with an ornithologist, Samira Agnihotri (henceforth SA) from the Indian Institute of Science, who has over 5 years' experience in observing, recording and identifying BRT birds in the company of Solega-speaking field assistants. Here, I first describe the patterns of Solega bird nomenclature, giving particular attention to which species are named, and which are not, and suggesting explanations for the differences. I believe that the latter is a particularly important question, one that is frequently ignored in studies on ethnobiological nomenclature. Next, I describe how the naming of birds in Solega is at times a dynamic, consensus-based process, where many of the phenomena described by Sillitoe [60] can be observed. Finally, I give an account of the very important role that birds play in Solega life, myth and ritual.

This chapter was previously published in the *Journal of Ethnobiology* as a paper co-authored with SA [167]. The following text contains an expanded 'Birds in Solega Life, Myth and Ritual', the main additions being texts and translations of bird-related myths. Additional data and analyses have also been added to Sect. 4.3.2 'Picture Elicitation Task' and Sect. 4.4 'Role of Perceptual Salience'.

4.2 Methods

A basic list of bird names in Solega was first compiled with the help of Solega-speaking field assistants from the ATREE field station (Fig. 4.1), where both authors were based. These were Solega men whose ages ranged from around 30 to 45, and included individuals who had worked with author SA in field situations while

A. Si, *The Traditional Ecological Knowledge of the Solega*, Ethnobiology,
DOI 10.1007/978-3-319-24681-9_4

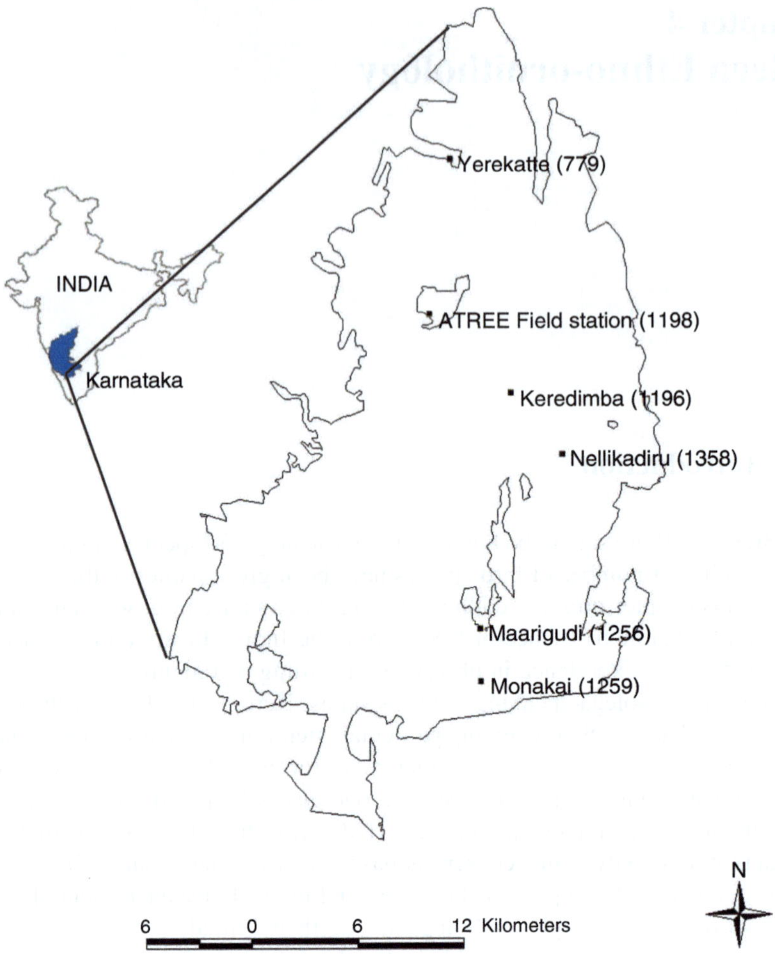

Fig. 4.1 Map of the BRT, showing location and elevation (in metres) of the field station and five settlements visited for the picture elicitation task. Previously published in [167]

collecting data for her research on bird behaviour and ecology. All field assistants lived in *po:ḍu* or villages in the vicinity of the field station and the Biligiri Rangaswamy Temple, which is frequented on a regular basis by largely Kannada-speaking tourists and devotees from the lowlands. As some of the bird names collected at this stage were bare mononomials, while some were binomials (of the form *x-hakki*, where *hakki* means 'bird'), the Solega field assistants were also asked to judge the acceptability of adding or removing the *hakki* morpheme, both when the name was used in isolation, or when it was embedded in a sentence, such as *a: marada me:le ondu x/x-hakki ku:tide* 'there is a x/x-bird sitting on that tree'. On the basis of these initial responses, the bird names were divided into three categories, namely 'obligate binomials', where the *–hakki* cannot be omitted, 'optional binomials', where the use of *–hakki* is optional, and 'obligate mononomials', where the addition of *–hakki* to the name is not allowed.

For the second stage of data collection, SA compiled a Powerpoint presentation consisting of the photos and calls of around 100 species of birds known to be present in the B. R. Hills. Five widely spaced Solega *poḍus* were visited (Fig. 4.1), and the presentation was shown to the inhabitants of those villages in a group—this ensured that the data arising from such elicitation sessions would represent a consensus opinion of the Solega living in a particular locality. Participants were shown the picture of a bird on a laptop screen, accompanied by the sound of the bird's call, and asked to provide an appropriate Solega name, if any. They were then asked to also relate any songs, stories or other items of folklore pertaining to that bird. As can be expected, there was considerable disagreement in the naming of some bird species, but in the vast majority of cases, a consensus was reached through discussion, and a single name was provided for a particular stimulus.

The names elicited in this way were assessed according to two criteria, namely reliability and consistency. The bird list compiled with the help of the field station Solega over several years of fieldwork was a useful benchmark against which to compare the responses from the villages, but we had to be open to the possibility that there might be variation in the names of certain birds from community to community. A lexical label for a particular species was therefore considered to be 'reliable' if it (a) matched the name for that species in our initial list, or (b) was novel, but was also accompanied by accurate information regarding the bird's behaviour, ecology or appearance. A name was considered 'consistent' if participants from three or more villages provided the same name (within reasonable limits of dialectal variation). It was possible for names to be both consistent and reliable (e.g. the responses from a majority of villages also matched our initial list), reliable, but not consistent (e.g. a novel label might be recorded in a single village, accompanied by a convincing amount of supporting information regarding the bird's biology), and consistent, but not reliable (e.g. a single, in our opinion, erroneous, name might be provided for a particular stimulus by participants from a number of villages). We might deem a particular response to be erroneous if the stimulus picture used in the task appeared to be misleading, i.e. if it drew attention to aspects of the bird's appearance that might not normally be visible in the wild. Such a case of mistaken identity would usually also be accompanied by incorrect biological information from otherwise knowledgeable speakers. Naturally, names could also be neither reliable nor consistent (e.g. participants might indulge in speculation when presented with an unfamiliar stimulus).

4.3 Solega Bird Nomenclature

It should first be pointed out that there are two types of naming phenomena in our data that can be labeled with the term 'binomial'. The first includes instances where a bird is named using the *x-hakki* form, as described above. It is in this sense that the term 'binomial' will be used in the remainder of the paper, i.e. at the 'generic' level, *sensu* [9]. The second type of phenomenon is the more familiar situation where a member of a generic class (a specific taxon, to use Berlin's terminology) incorporates within its name the label for that class as a whole. A Solega example from our data is *elesiṭṭe* 'tailorbird', which is placed within the superordinate category *siṭṭe* 'small passerines'.

4.3.1 Acceptability Judgements

Based on the initial acceptability judgements of the field assistants, it was found that only a handful of birds are named with obligate binomial lexemes, where *-(h)akki* is an inalienable part of the name (Table 4.1). An intermediate number of birds are named by optional binomials, where the *-(h)akki* can be used or dropped, at the speaker's volition. The majority of bird names are obligate mononomials, where the addition of *–(h)akki* is not allowed. A cursory scan of Table 3.1 might appear to confirm some 'universals' of ethnoclassification proposed by Berlin [9]. These include (1) the dominance of mononomials among what Berlin calls generic taxa, (2) a tendency for 'perceptually salient' organisms and monotypic genera to be preferentially named, and (3) a systematic overlap between 'folk' and 'scientific' taxa, especially at the level of the folk genus.

The first of these generalisations is the most readily supported by our data. Overall, a majority of bird names (35 out of 55, or 64 %) are obligate mononomials. However, the use of the 16 (29 %) 'optional' species names in their full, binomial forms is by no means marginal—the linguistic context plays a crucial role here. When asked "What bird is that?" a Solega person might give the following reply:

(1) *adu e:va hakki?*
that which bird
'What bird is that?'

(2) *adu ondu sore*
that one dove
"That's a dove."

However, if the same speaker intended to draw a listener's attention to the presence of a particular bird, s/he might spontaneously say the following:

(3) *no:ḍ-u alli ondu sore-yakki ku:t-ide*
look-IMP—there one dove-bird sit-PERF
"Look, there's a dove sitting there."

When asked, Solega speakers will undoubtedly say that it is acceptable to omit *–(y)akki* from sentence (3), but the majority of their spontaneous utterances contain the superordinate taxon. This has important implications for the way in which data are collected in ethnobiological studies. Traditional question and answer sessions where a researcher repeatedly asks a consultant to name dead or living specimens,

Table 4.1 Named birds in Solega, along with their official English and scientific names

Solega	English	Scientific
Obligate binomials (6)		
araḷakki	Emerald Dove	_Chalcophaps indica#_
maṇṇakki*	Crested Treeswift	_Hemiprocne coronata#_
	House Swift	_Apus affinis#_
	Asian Palm Swift	_Cypsiurus balasiensis_
	Barn Swallow	_Hirundo rustica_
ma:diyakki*	Scarlet Minivet	_Pericrocotus flammeus_
sivanakki	Greater Flameback	_Chrysocolaptes lucidus#_
gumisiyakki*	House Sparrow	_Passer domesticus#_
to:ḷiyakki	bats	
beḷḷa to:ḷi	?	
a:regã to:ḷi	?	
suṇḍã to:ḷi	Indian Pipistrelle	_Pipistrellus coromandra_
Optional binomials (16)		
ba:sa:tale(hakki) or _taleba:sã_	Brahminy Starling	_Sturnus pagodarum_
piggã(nakki)	Forest Wagtail	_Motacilla indica_
	Grey Wagtail	_M. cinerea_
	White-browed Wagtail	_M. maderaspatensis_
ṭhĩya(nakki)	Red-wattled Lapwing	_Vanellus indicus_
kuṭru/kuṭrakki	White-cheeked Barbet	_Megalaima viridis_
	Brown-headed Barbet	_M. zeylanica_
	Crimson-fronted Barbet	_M. rubricapilla_
kambarakuṭru*	Coppersmith Barbet	_M. haemacephala_
ke:sã/ke:sanakki^	Indian Cuckoo	_Cuculus micropterus#_
gi:jiga(nakki)	Baya Weaver	_Ploceus philippinus_
si:pi(yakki)*	Indian Silverbill	_Lonchura malabarica_
	Black-headed Munia	_L. malacca_
	Scaly-breasted Munia	_L. punctulata_
	Common Rosefinch	_Carpodacus erythrinus#_
sore(yakki)	Spotted Dove	_Streptopelia chinensis_
	Eurasian Collared Dove	_S. decacto_
araḍe(y-akki)	Rufous Treepie	_Dendrocitta vagabunda_
uṇṇigirika(n-akki)	Common Myna	_Acridotheres tristis_
	Jungle Myna	_A. fuscus_
	Bank Myna	_A. ginginianus_
oggeguḍukã(n-akki)	Indian Roller	_Coracias benghalensis#_
karaḷi(y-akki)	Black Drongo	_Dicrurus macrocercus_
	Bronzed Drongo	_D. seneus_
	White-bellied Drongo	_D. caerulescens_
doḍḍakaraḷi	Greater Racket-tailed Drongo	_D. paradiseus_
pa:riva:ḷa(d-akki)	Blue Rock Pigeon	_Columba livia_

(continued)

Table 4.1 (continued)

Solega	English	Scientific
morã(n-akki)	Blue-Winged Parakeet	*Psittacula columboides*
gi:ṇa(n-akki)	Rose-Ringed Parakeet	*Psittacula krameri*
saṇṇa gi:ṇa	Vernal Hanging Parrot	*Loriculus vernalis#*
se:na(n-akki)	Hill Myna	*Gracula religiosa#*
Obligate mononomials (35)		
ko:ḷi	Domestic fowl	*Gallus gallus*
ka:ḍako:ḷi	Grey Junglefowl	*G. sonneratii*
*kapuṭa^*	Grey Nightjar	*Caprimulgus indicus*
	Large-tailed Nightjar	*C. macrurus*
	Indian Nightjar	*C. asiaticus*
ka:kanakoṭe	Indian Scimitar Babbler	*Pomatorhinus horsfieldii#*
ka:ge	Large-billed Crow	*Corvus macrorhynchos*
ni:ruka:ge	Indian Cormorant	*Phalacrocolax fuscicollis*
ka:nakattale	Black Eagle	*Ictinaetus malayensis#*
ka:nagoravã	Malabar Whistling Thrush	*Myiophonus horsfieldii#*
*kuggilu^*	Mountain Imperial Pigeon	*Ducula badia*
	Green Imperial Pigeon	*D. aenea*
	Nilgiri Wood Pigeon	*Columba elphinstonii*
bu:jore	Laughing Dove	*Streptopelia senegalensis*
kembuttã	Greater Coucal	*Centropus chinensis#*
keshṭe	Red Spurfowl	*Galloperdix spadicea*
kokkore	Migratory waterbirds	
ko:gile	Asian Koel	*Eudynamys scolopacea#*
(ko: ḷi)koṭã	Oriental Honey Buzzard	*Pernis ptilorhyncus#*
garuḍa	Brahminy Kite	*Haliastur indus#*
giḍugã	Common Kestrel	*Falco tinnunculus*
	Black-shouldered Kite	*Elanus caeruleus#*
gi:ṇa	Plum-headed Parakeet	*Psittacula cyanocephala*
*gu:be^*	Oriental Scops Owl	*Otus sunia*
	Collared Scops Owl	*O. bakkamoena*
gumma	Brown Hawk-owl	*Ninox scutulata#*
	Brown Fish Owl	*Ketupa zeylonensis#*
	Spot-bellied Eagle-owl	*Bubo nipalensis#*
goregosha:kã	White-throated Kingfisher	*Halcyon smyrnensis*
navilu	Indian Peafowl	*Pavo cristatus#*
*natta:re^*	Jungle Owlet	*Glaucidium radiatum#*
*siṭṭe**	small passerines	
*elesiṭṭe**	Common Tailorbird	*Orthotomus sutorius#*
*karisiṭṭe**	Indian Robin	*Saxicoloides fulicata#*
*koḍangisiṭṭe**	White-browed Fantail	*Rhipidura aureola*
*tu:gusiṭṭe**	Purple Sunbird	*Nectarinia asiatica*
	Purple-rumped Sunbird	*N. zeylonica*

(continued)

Table 4.1 (continued)

Solega	English	Scientific
	Crimson-backed Sunbird	*N. minima*
	Loten's Sunbird	*N. lotenia*
*biḷisiṭṭe**	Oriental Magpie-Robin	*Copsychus saularis*
*pa:pira^*	Common Hawk-cuckoo	*Hierococcyx varius*
*maratottã**	Velvet-fronted Nuthatch	*Sitta frontalis*
	Chestnut-bellied Nuthatch	*S. castanea*
mara kuṭuka	Rufous Woodpecker	*Celeus brachyurus#*
	Streak-throated Woodpecker	*Picus xanthopygaeus*
	Yellow-crowned Woodpecker	*Dendrocops mahrattensis*
maragosappa:na	Black-rumped Flameback	*Dinopium benghalense*
moḍemoka	Hoopoe	*Upupa epops*
sha:ḷugã/vijju	Besra	*Accipiter virgatus*
	Shikra	*A. badius*
*saṭṭugaba:la**	Asian Paradise Flycatcher	*Terpsiphone paradisi#*
*saṇeyã**	Green Bee-eater	*Merops orientalis*
	Chestnut-headed Bee-eater	*M. leschenaultii*
	Bluebearded Bee-eater	*Nyctyornis athertoni#*
si:danakari	Jungle Babbler	*Turdoides striatus*
doḍḍa si:danakari	Rufous Babbler	*T. subrufus*
hullu si:danakari	Yellow-billed Babbler	*T. affinis*
	Tawny-bellied Babbler	*Dumetia hyperythra#*
haddu	Raptors	
hebbaddu	Unknown	
biliyaddu	Unknown	
koṭrole	Red-vented Bulbul	*Pycnonotus cafer*
koṭṭipiḍiyã	Red-whiskered Bulbul	*P. jocosus*
*maḷegoḍḍã^*	Orange-headed Thrush	*Zoothera citrina*

The numbers in parentheses indicate the number of 'generic' names per group. The left-aligned names in column 2 are 'generic' names *sensu* [9]; the right-aligned names are 'subgeneric' or 'specific'. Superscripts roughly indicate those birds which are not visually striking—* small; ^ cryptic; # after the scientific name indicates a locally monotypic genus. The scientific identifications are from [168]

either in situ or in a sorting task, would, in languages like Solega, clearly bias speakers' responses towards one type of label, namely the bare mononomial. It was noted in Chap. 2 that Solega plant names behave more like obligate binomials, where words like *mara* 'tree' are dropped only in the context of a person reciting a list of trees; in the case of grasses and vines, the morphemes for 'grass' and 'vine' are totally obligatory. The fact that a context has been unambiguously established (along the lines of 'We are now going to talk about birds') has a very real effect on subsequent utterances, and so it is worth considering whether the data obtained from such elicitation sessions is indeed 'naturalistic'. To use an English analogy, a car enthusiast might well refer to his/her prized possession as an 'E-type', but only in the limited contexts of addressing a fellow enthusiast, or continuing a discussion on Jaguars. Assuming, then, that the optional mononomials in Solega are more routinely used as binomials in spontaneous utterances (see the following section for further discussion), the total proportion of binomials at the 'generic' level comes to 38 %—still a minority, but a sizeable one. One prediction of Berlin that does hold unequivocally is that Solega bird names have very few subordinate or 'specific' taxa that are labelled by secondary lexemes.

4.3.2 *Picture Elicitation Task*

We recognise that the elicitation of bird names from static, two-dimensional images, albeit accompanied by recordings of the birds' calls, is a poor substitute for sighting the bird in its natural habitat in the company of a Solega speaker (see [169, 170] for further commentary). For that reason, the following discussion is based on only responses that were assessed as either 'reliable and consistent' or at least 'reliable' (see Methods section for definitions). The results of the picture elicitation task carried out in five villages provide further support for our claim that in Solega at least, the preponderance of mononomials in speakers' responses is likely to be an artefact of the data-gathering process. The effect of context on the naming and classificatory responses of Nuaulu speakers has already been discussed in detail by Ellen [59] for two culturally significant organisms—the cassowary and the cuscus. Criticizing theory-driven ethnobiological studies for their "single-mindedness", their "failure to include a sufficiently critical assessment of the context and variation of the data elicited" and the "spurious uniformity" of their data, Ellen stresses that:

> ...classifications are not ossified rubrics of cerebral apparatus, genetically programmed, although the principles that order them may be; they are culturally-determined ordering devices interacting constantly with experience, infinitely variable both over time and between individuals. (p. 202)

In this study, a basic quantitative analysis of the pooled responses from all five villages shows that the majority of first responses (i.e. the first name provided by a participant on viewing the picture of a bird and hearing its call) were bare mononomials (69 % of a total of 237 tokens). This is unsurprising, given the artificial nature of the task, and the constraints this might impose on the participants' responses. However, an interesting phenomenon evident in this data set is the reversal of a participant's first response into its opposing form, e.g. a name is first provided in its mononomial form, and repeated soon afterwards—by the same speaker, or by another—in a binomial form, which incorporates *–hakki*.

Speaker 1: *adu araḍe* <PAUSE> *araḍehakki*
'That's a treepie. <PAUSE> Treepie-bird.

It might seem at first that such 'reversals' are merely indicative of free variation between the two forms of the optional binomials presented in Table 4.1. After all, such reversals only occurred in a minority of situations (around 27 %). However, a closer look at the frequencies of the two types of possible reversal reveals a distinct pattern: reversals from bare mononomial to binomial, as in the example given above, occurred 66 % of the time (42 tokens), while reversals in the opposite direction occurred in only 34 % of instances (22 tokens). The fact that speakers are more likely to replace bare mononomials with binomials, than vice versa, suggests that the former is perceived as being incomplete in some way. My observation, that binomials are followed up with mononomials far less frequently, indicates that the two forms are not in free variation.

These findings relate a cautionary tale: it is unwise to assume that the utterances recorded in controlled elicitation tasks are totally 'naturalistic', simply because the questions were asked in the target language, and the majority of speakers responded in a particular way. In Solega at least, the naming of birds seems to be quite context dependant, and optional binomials often appear in their longer form in connected speech. In the following extract, the Solega consultant had been speaking for the last 10 minutes or so about the various birds found in his forest, and decided, quite spontaneously, to start talking about *kuruḷi-hakki* 'Common Quail'. After a brief comment about how rare this species has become, he goes on to relate a folk tale that explains why the elephant has no scrotum (and why, incidentally, it has two large bumps on its head). Here, the transition from the previous narrative to the new species is marked by a bare mononomial, *kuruḷi*, but in all subsequent tokens, the name of the bird appears as a binomial. The speech presented in the following passage is arguably far more naturalistic than that obtained from the picture elicitation task, and strongly supports our view that when the morpheme *–hakki* is optional, the bare mononomial is actually the marked form, and the longer, binomial form the more natural choice.

> ***Kuruḷi*** *…* ***kuruḷi*** *hatt hattu ip-ippattu iddõ. A:* ***kuruḷi-hakki*** *endale saṇṇa hakki. Tanna hesarave 'kuruḷi! kuruḷi! kuruḷi!' enduru koṭṭade. "O:* ***kuruḷi-hakki*** *ha:ḍdade." A:* ***kuruḷi-hakki*** *e:na ma:ḍi kittu, a:nette oitittu … a:ne ho:ga timeinalli e:n a:gottu prrr-enna a:ri kittu! A:nega a:rukku endottu. A:ne hi:ḍutte keḷage ittu … ade:na:gottu, prrr-enna a:radakka ue adara hi:ḍutte me:le ku:turtu. A:* ***kuruḷi-hakki*** *hi:ḍa e:rusuṭṭattu. A: hi:ḍu keḷage ille me:le ade, katti nalli. A:* ***kuruḷi-hakki*** *o:ḍsuṭṭattu. Gaṭa no:ḍidale ba:ri gaṭa,* ***kuruḷi-hakki*** *ira:du ipaṭiye. A: tara ma:ḍittu,* ***kuruḷi-hakki****.*
>
> Quails … there used to be quails in groups of 10 or 20. The quail is a small bird. It speaks its own name, *'kuruḷi! kuruḷi! kuruḷi!'* "Oh, I can hear quails chirping," (one would say.) What did the quail once do? An elephant was walking along … and at that moment, a quail flew up, 'PRRR!' The elephant got a fright. The elephant's balls used to be down there … but then, when the bird flew up, 'PRRR!' the elephant's balls jumped up. The quail had raised its balls up. The balls aren't down there, they're sitting on top. The quail had lifted them up. If you think about it, an elephant's really big, the quail's only little. But that's what it did, the quail.

An unexpected result that emerged from the picture elicitation task was the variation, from village to village, in the membership of the categories 'obligate binomial', 'optional binomial' and 'obligate mononomial'. Although speakers from these villages were not systematically asked to state their preference for one form over the other, several of their responses were found to be different from the acceptability judgements we had earlier recorded with the Solega field assistants from nearby settlements. In particular, we were surprised to find that many of the 'obligate

Table 4.2 Selected bird names from the picture elicitation task

Optional binomials	
Alternative Solega name	Common name
keshṭe hakki	Red Spurfowl
marakuṭuka hakki	Woodpeckers
gi:ṇanakki	Plum-headed Parakeet
goravana hakki	Malabar Whistling Thrush
siṭṭe hakki	Small passerines
ko:gilakki	Asian Koel
koṭrole hakki	Red-vented Bulbul
kuggilu hakki	Pigeons
pa:piranakki	Common Hawk-cuckoo
si:danagari hakki	Jungle Babbler
gumusi hakki	House Sparrow

New obligate binomials	
Solega name	Common name
arsinakki	Yellow-browed Bulbul (*Acritillas indica*)?
ka:rihakki	Woodpeckers
ku:sakki, nana ku:so hakki	Puff-throated Babbler
ka:raȳyanakki	Woodpeckers
mara eḷe hakki	Nuthatches
sa:vakki	Flameback woodpeckers

'Optional binomials' are those bird names from Table 4.1 which were initially considered to be obligate mononomials. The only exception is 'sparrow', which was initially thought to be an obligate binomial. 'New obligate binomials' are some newly-recorded bird names that were absent from the initial list

mononomials' from our initial name list were in fact optional binomials, at least for speakers from some villages. Table 4.2 gives a list of obligate mononomials from Table 4.1, which, during the course of the picture elicitation task, were used in conjunction with the morpheme *–hakki*.

Some new obligate binomials were also recorded during the picture elicitation task (Table 4.2). Scrutiny of the morphemes that make up such names shows that binomial names are obligatory when the first morpheme encodes an attribute of the bird being named—this can include physical appearance (e.g. *arsinakki* 'yellow bird'), behaviour (*mara eḷe hakki* 'tree climb bird'), cultural significance (*sa:vakki* 'death bird'), an associated object, substance or plant (*maṇṇakki* 'mud bird') or a god's name (*ka:raȳyanakki* '*karaȳya*'s bird). Taking the new data in Table 3.2 into account, the revised counts and percentages for the different types of names are as follows:

Obligate binomials: 12 (19 %)
Optional binomials: 27 (44 %)
Obligate mononomials: 23 (37 %)

Binomials (totalling 63 %) now outnumber the obligate mononomials, and it is very likely that at least some of the latter category could also be used in binomial form. However, this remains to be confirmed, as many of the bird names from our initial list (Table 4.1) were not recorded in the picture elicitation task.

The picture elicitation task also revealed the ease with which new names could be coined for yet unnamed birds, or birds with which a speaker might be unfamiliar. Some of these are given in Table 4.3. In one village, a speaker took great delight in composing novel names for some birds that he was unable to identify—such attempts were greeted with much amusement from the other Solega speakers participating in the task, and indicated to us that the names being provided were indeed nonce creations. Names like *piṭre hakki* and *pigganakki* were offered for a couple of small birds—in Solega, *piṭre* refers to something that is small and useless, while *piggã* means a young child who runs around restlessly, and does not listen to the admonishments of his/her elders. Similarly *guggalakki* (*guggalu* means 'solitary' or 'loner') could indicate a bird that is shy of humans, and is usually encountered by itself in the forest. Such names may well be the product of a single speaker's imagination, but could very easily become established in the lexicon of that speaker's village dialect, if s/he were to use it repeatedly and consistently for a particular referent (i.e. a particular species or group of species). Therefore, while it is tempting to

Table 4.3 Solega bird names for which reliable identifications cannot be provided

Solega name	Possible identification
*da:sapaṭada hakki**	Scimitar Babbler?
goregotã	Kingfisher? bee-eater?
goresubbã	White-throated Kingfisher
gu:ge	
gubuḷegaṇṇa	Owl? Oriental White-eye?
guḍugaḍe	Spurfowl?
*guggalakki**	Laughing Dove
goujalakki	Type of quail/francolin?
guḍuga:ḍe hakki	Type of quail/francolin?
*ka:nakki**	'Evergreen forest bird'
*ka:nuguḍḍã**	
ke:sehakki	
kuigalakki	
kusurihakki	
ko:kunji	
meṇḍegaṇa hakki	'Large-eyed bird'
moḍevaḷakki	
moṭṭesillã	
ratna giṇi	Vernal Hanging Parrot
soĩna gi:ṇa	Vernal Hanging Parrot
solegitti hakki	Paradise Flycatcher?
taragaḍakã	Orange-headed Thrush

*After the Solega name indicates a probable nonce coining. ? indicates a high level of doubt

dismiss such names as *arsinakki* 'yellow bird', *ka:nakki* 'evergreen forest bird' *kuĩ kuĩ hakki* 'bird that calls *kuĩ kuĩ*' and *koṭṭi hakki* 'crest bird' as nonce coinings, as they were only recorded in one village, it is important to remember that such names are, in essence, not dissimilar to names such as *araḷakki* 'castor bird', *kuṭrakki* 'bird that calls *kuṭru*', *ka:rihakki* 'calling bird' and *maṇṇakki* 'mud bird', which were obtained from several villages. Quite often, several speakers from a single village would agree that a name like *arsinakki* 'yellow bird' was the correct name for a particular species of yellow bird; to us, this indicated that *arsinakki* was the accepted name for that bird for the speakers of that village, even though our consultants from the field station might argue that such a name did not exist in their lexicon.

A related issue is that of variation, of which there was a significant amount in the naming of some species of bird. The level of variation ranged from minor differences in pronunciation to completely novel names, as shown in Table 4.4. Some of the variation, not shown in 4.4, but occurring frequently in our conversations with Solega consultants, could be attributed to pragmatic factors, such as the inability to satisfactorily identify a bird in the wild, due to its distance from the speaker, or its being obscured by foliage, for instance. In such instances, a common response would be that the bird was a kind of *siṭṭe* 'small passerine'. Many small birds, that are not known to have formal Solega names were also identified in this manner. This is identical to the situation described by Descola (1994; cited in [60]) for the Achuar Jivaro of the Peruvian Amazon, who use the word *chinki* 'sparrow' under similar circumstances. According to Sillitoe, this could be a strategy to "facilitate agreement over the naming of animals, seen fleetingly, for example, when precise identification is difficult".

Much of the variation shown in Table 4.4 is small enough to be restricted to the domain of phonological dialectal variation, but for some species, such as the coppersmith barbet, the bulbuls and the sunbirds, the names indicate a significantly different conception of the same bird by two or more communities that otherwise share a common language. The label *kammarã kuṭru* (lit. blacksmith barbet), which seems curiously similar to the common English name of this bird 'Coppersmith Barbet', was recorded from two villages, and makes reference to the call of this bird, which sounds like a smith's hammer at work. In one village, however, the name *batta kuṭru* (paddy *kuṭru*) was recorded, and here the participants explained that the name alluded to the resemblance between the bird's call and the sound produced by the action of dehusking rice. Thus, the distinctive call of the bird was still being used as the identifying modifier, but in a subtly different way. In the case of the bulbuls, the names *koṭrole* 'Red-Vented Bulbul' and *koṭṭipiḍiyã* 'Red-Whiskered Bulbul' were recorded from two villages (the same names were provided by the Solega men working at the field station), but speakers from two other villages grouped these birds together, saying that one was a male (*gaṇḍu koṭṭi*, Red-Whiskered Bulbul), and that the other was a female (*heṇṇu koṭṭi*, Red-Vented Bulbul).

A further case of lexical variation is that of *tu:gusiṭṭe* 'hanging bird', which was the label provided for the four sunbird species found at our field site, both by the Solega men working at the field station, and by picture task participants from one village. The name refers to the hanging nests that are constructed swinging from

Table 4.4 Variation in bird names across villages—data from the picture elicitation task

	Keredimba	Nellikadiru	Ma:riguḍi	Moṇakai	A:repa:ḷya
Red Spurfowl	*(ka:ḍa)kesaṭe*	*keshṭe*		*keshṭe*	
Rufous Woodpecker		*marakuṭuka hakki* 'woodpecker'	*ka:rañyana hakki* 'Ka:rañya's bird'	*ka:ri hakki* 'calling bird'	*sa:vakki* 'death bird'
Flameback woodpeckers	*sa:vu hakki* 'death bird'	*ka:ri hakki* 'calling bird'	*ka:ri hakki* 'calling bird'	*ka:ri hakki* 'calling bird'	*sa:vakki* 'death bird'
Coppersmith Barbet	*kammara kuṭru* 'blacksmith *kuṭru*'	*kammara kuṭru* 'blacksmith *kuṭru*'	*batta kuṭru* 'paddy *kuṭru*'		
Indian Cuckoo	*ke:sakki*	*ke:sanakki*	*ke:sã*	*ke:sanakki*	*ke:sakki*
Greater Coucal	*kembuti/kembuguti*	*kembuta*	*kembuta*	*kembuta*	
Brown Fish Owl	*gumma*	*gumma*	*gumma*	*gumma*	*gubuḷegaṇṇa*
Spotted Dove	*totore*	*sore*	*male sore* 'mountain *sore*'	*sorehakki*	*sorehakki*
Laughing Dove	*bu:jore*	*sore*	*na:ḍa sore*	*guggalakki* 'solitary bird'	*bu:jore*
Black Eagle	*ka:nakatale* 'darkness of the forest'	*hebbaddu* 'big raptor', *kuruḷaddu* 'quail raptor'	*ka:nakatale* 'darkness of the forest'		
Red Wattled Lapwing	*(kuṇḍu)biḷuka*	*ṭhĩyã*	*ti:ṭrigã*	*ti:ṭigã*	*ṭĩya:nakki*
Rufous Treepie	*araḍe*	*araḍe hakki*	*araḍe*	*moḍeva:ḷakki, kurubigo:ḍi*	*araḍe*
Scarlet Minivet	*ma:di hakki*	*ma:di hakki*	*ma:di hakki*	*ma:dilakki*	
Asian Paradise Flycatcher	*saṭṭugaba:la* 'ladle tail', *asagã*		*solegitti hakki*	*ma:dilammi hakki*	
nuthatches	*mara eḷe hakki* 'tree-climbing bird'	*maratotta* 'tree jumper'	*maratotta* 'tree jumper'	*maratotta* 'tree jumper'	*maratonta* 'tree jumper'

(continued)

Table 4.4 (continued)

	Keredimba	Nellikadiru	Ma:riguḍi	Moṇakai	A:repa:ḷya
Ved-vented Bulbul	*koṭrole*	*koṭrole*	*heṇṇu koṭṭi* 'female *koṭṭi*'	*heṇṇu koṭṭi* 'female *koṭṭi*'	
Red-whiskered Bulbul	*koṭṭipiḍiyã*	*koṭṭipiḍiyã*	*gaṇḍu koṭṭi* 'male *koṭṭi*'	*gaṇḍu koṭṭi* 'male *koṭṭi*'	
Jungle Babbler	*si:danagari*	*doḍḍa si:dagari* 'big *si:danakari*'	*si:dari*	*si:dari*	*si:danagari*
Puff-throated Babbler		*nanna ku:so* 'my child'	*ku:sakki* 'child bird'	*nanna ku:so hakki* 'my child bird'	
wagtails	*piggã*	*ni:rpiggã* 'water *piggã*'	*pigganakki*	*ni:rpikka* 'water *pikka*'	*pigganakki*
sunbirds	*tu:gusiṭṭe* 'hanging small passerine'		*hu: karaḷi* 'flower drongo'	*kokkakki* 'beak bird'	

Only those species with consistent identifications from three or more villages are shown here. The exceptions are the laughing dove, for which two names were obtained from two villages each, and the sunbirds, one of whose names matches that from our initial list small branches. In another village, however, the label *hu: karaḷi* 'flower drongo' was provided. We assessed this as reliable, because the participants agreed on this name unanimously, and were also able to provide details of the physical characteristics (small size, bright colours), behaviour (sucking nectar from flowers) and distribution (high-altitude forests) of these birds. In a third village, the name *kokkakki* was recorded, a clear reference to the birds' long, curved beaks. It seems that while the former group of Solega speakers seemed content to place these birds in the superordinate category *siṭṭe* 'small passerines', the second grouped them together with *karaḷi* 'drongos', possibly owing to their rapid, darting flight and the metallic sheen of their feathers. The final group, in contrast, provided a name that did not affiliate these birds with either small passerines or with drongos.

A final example of variation that is worth discussing at this point is the case of the puff-throated babbler, a culturally-important bird whose call is said to be *nanna ku:so* 'my child!' There is a universally-known folk tale in Solega that explains why this bird came to lament a lost child (see below), and every single speaker who heard the recording of this bird's call during the picture task was able to recognise it as the bird that had lost its child. Surprisingly, a great number of people who know the story, and who can correctly identify the bird call, are unable to provide a name for it—while some say that they simply do not know a name for it, others reply, with obvious hesitation, that they call it the *nanna ku:so hakki* ('my child' bird), and a only small minority provides the far more compact label *ku:sakki* 'child bird'.

4.4 The Role of Perceptual Salience

It is almost a tautology to state that human perception will have a significant influence on a language community's folk classification system—indeed, it is hard to imagine any form of human behaviour that is not similarly influenced. However, it would be unwise to conclude that humans merely react to the input from their sense organs to automatically produce mental (and by extension, lexically encoded) representations of the natural world (see [171] for a detailed discussion on this topic). Despite decades of comparative linguistic research on various domains of semantics, the number of true cross-linguistic semantic universals that have withstood rigorous empirical testing remain few in number [172]. It is therefore to be expected that different languages will also 'carve up' the natural world in different ways, giving rise to a range of folk classifications. As previously suggested by other authors (e.g. [15, 62, 153]), cultural or utilitarian factors should also play an important role in determining which organisms are to be named, and how distinctive those names should be.

In keeping with the hypothesis of perceptual salience, many of the large, visually striking birds in the B. R. Hills do have Solega names. This is only part of the explanation, however, as a more detailed investigation into which birds are named, and which are not, reveals a far more complicated pattern. For instance, it is not unusual for ethno-ornithologists to report that their target languages have distinct labels for birds that can be described as 'rare and similar-appearing sibling species', 'small and dull-coloured' and 'obscure' [169]. First, however, it is pertinent to clearly establish what is meant by 'perceptually salient', as it is all too easy to explain away inconvenient facts which do not fit theoretical models by means of vague labels such as 'salient' and 'not salient', 'similar' and 'not similar'. It is probably not possible to devise a single, objective, cross-linguistic and cross-taxon criterion of 'salience', but we could, at the very least, settle on a minimum salience threshold for a taxon in a single language. Two birds that are given distinct names but share many visual characteristics are *koṭṭipiḍiyã* (the Red-Whiskered Bulbul, *Pycnonotus jocosus*) and *koṭṭorole* (the Red-Vented Bulbul, *Pycnonotus cafer*). Apart from having a *koṭṭi* 'crest' on their head, both birds are roughly the same size, and the same colour overall. The principal differences are that the former has two small red 'whiskers' and a whiter face, while the latter has a darker head, and a patch of red under its tail. Clearly, these differences are sufficient to warrant the naming of these birds by two different labels. The owls are another set of birds that are contrasted by few features—*gumma* is the name given to three large owl species, while two smaller species of Scops Owl are called *gu:be*. A difference in size, then, can be a perceptually salient feature of Solega bird taxonomy. At the generic/specific interface, three species of birds, *si:danagari* the Jungle Babbler, *doḍḍa si:danagari* the Rufous Babbler and *hullu si:danagari* the Yellow-Billed Babbler, are recognised as distinct folk species that nevertheless belong together. All three species of birds move in noisy flocks, and the only distinguishing visual features seem to be differences in their size and overall colour—which may be described as drab grey, dark olive and grey, respectively. The magnitude of the difference, while not remarkable, is clearly sufficient for these birds to be given related, but different names.

Given that visual cues as minor as the ones discussed above can be associated with differently named birds, it is surprising that three species of woodpecker are named *marakuṭuka* when the differences between these species are at least as prominent as those between the bulbuls. In a similar vein, three species of mynah, which appear at least as different as the babblers, are all called *uṇṇigirika:nakki*, while four species of drongo, distinguishable even from a distance by differences in body size, colouration and the shape of their ornamental tail feathers, are all called *karaḷiyakki*. The only drongo that has a unique, 'specific' name is the Greater Racket-Tailed Drongo, which is labelled *doḍḍakaraḷi* 'big *karaḷi*'. Much cultural significance is attached to this bird, however (see below), and perceptual reasons alone (its larger size and trailing tail feathers) cannot account for its special nomenclatural status.

Some other phenomena arising from the picture elicitation task cannot be explained by the perceptual salience hypothesis alone. Three species of orioles—bright yellow birds the size of mynas with prominent calls—live in the B. R. Hills, but no one was able to provide a name for any of them. In contrast, the significantly smaller *ma:di hakki* 'Scarlet Minivet', which is similarly coloured, was correctly named in 4 out of 5 villages. Most participants were even able to point out that only the female was bright yellow, while the male had scarlet plumage. Another bird that no Solega speaker was able to name is the Yellow-Footed Green Pigeon (*Treron phoenicoptera*). The absence of the name of this bird from the Solega lexicon is curious, as practically all other members of the family Columbidae are easily recognised and named—these include *kuggilu, sore hakki, bu:jore, pa:riva:ḷa* and *araḷakki* (Table 4.1; also see the section below on 'Relation to scientific taxonomy). Even more inexplicable, however, was our observation that not a single participant tried to insist that this obviously pigeon-like bird was really just an example of one of the other five columbid taxa. Some reported that they had seen the bird before, but that it simply did not have a name. Most English speakers, when faced with a similar predicament, would probably not shy away from stating that *T. phoenicoptera* was 'a kind of pigeon', in much the same way that they would unhesitatingly classify a hitherto unseen breed of dog as 'a kind of dog'. No Solega speaker, however, spontaneously categorised *T. phoenicoptera* as 'a kind of *kuggilu*' or 'a kind of *sore hakki*'. This indicates that Solega does not have a pigeon/dove/Columbid umbrella term or 'covert category' which could be posited for English speakers.

When speakers are explicitly asked to mention the morphological features used as diagnostics for identifying different birds, they may not always agree on the salient feature(s) to be used. In the case of the parrots/parakeets, speakers are often divided on the issue of which morphological criteria should be used to distinguish *mo:rã* from *gi:ṇa*. Ultimately, however, there seems to be a consensus that the blue-winged parakeet is *mo:rã*, while the rose-ringed parakeet and the blossom-headed parakeet are both *giṇa*.

> *Mo:rãdu kempu kokku, mai ondu tarada ni:li baṇṇa, kattinalli kari da:ra. Gi:ṇa endale mai asiru pu:ra, arsina kokku. <Jaḍegowḍa, Bangalipo:ḍu>*
> *Mo:rã* has a red beak, and the body is sort of blue, with a black line on the neck. *Gi:ṇa* has a completely green body, and a yellow beak.
>
> *Gi:ṇana pari:kshe ma:ḍagiddale na:vu, hakki hi:ge ku:tira:ga, illi ondu boṭṭu ade. Balasari ondu boṭṭu, rekke me:le, puna eḍasari ondu boṭṭu irtade. Mo:rãna hakkili adu baradille. <Da:segowḍa, Bu:ta:nipo:ḍu>*
> To distinguish *gi:ṇa*, when the bird is sitting like this, there is a spot here. [POINTS TO RIGHT ARM] There is a spot on the right, on the wing, and another one on the left. The *mo:rã* does not have these.
>
> *Kempu taledu gi:ṇa. <Madegowḍa, Nellikadiru po:ḍu>*
> The one with a red head is *gi:ṇa*.

The reason for this disagreement cannot be a lack of perceptual salience, as all three species are commonly seen and heard throughout the year in the B. R. Hills. Moreover, both *gi:ṇa* and *mo:rã* are significant crop pests, and can completely lay waste to a *ra:gi* 'finger millet' field that is left unwatched. Instead, as Sillitoe suggests [60], the answer may lie in the possibility that we had asked our consultants an inappropriate question in the first place.

> When asked how they identify particular animals, informants vary in their responses. They are not accustomed to specifying what cues they look for as naming criteria... They see any creature as a distinctive whole, considering simultaneously a range of observable cues, not seeking a few characteristic ones. (p. 1169)

As stated earlier, most of the large and easily spotted birds in BRT do have Solega names, but as Table 4.1 shows, a significant number (15 out of 55 at the 'generic' level) are either small or cryptic species. Similarly, a significant number of birds found in this locality, that are either medium to large sized, visually striking or common, or that possess distinctive calls, lack any Solega name whatsoever (Table 4.5). The presence of many of the visual cues that were arguably 'perceptually salient', in differentiating various named species cannot therefore guarantee that species possessing those features will be named. Very large and visually distinctive migratory waterbirds that appear in BRT during the winter months (December to February) are all lumped together in the category *kokkore*, and the Solega make no attempt to try and distinguish between, say storks and herons. It could be argued that such species, being transient, do not leave as great an impression on the Solega psyche as birds that are present all year round. Such an argument can, however, be easily countered by the observation that many cryptic birds such as *ke:sanakki* 'Indian Cuckoo' and other cuckoos are primarily recognised by their calls. *Ke:sanakki* is particularly significant in this respect, because its calls are usually only heard in April–May, and herald a change of season; for the rest of the year, this species is rarely seen or heard.

A strong claim made by Berlin [9] relates to the special status of monotypic genera. In this regard, Berlin says that monotypic species stand out in the biological landscape due to their relative evolutionary isolation, and are consequently more likely to be given a distinct name in an ethnoclassification scheme for being perceptually more salient.

> Focusing solely on monotypic genera as the most likely candidates for linguistic recognition, the following hypothesis is suggested:
>
> 1. If a scientific genus, x, is monotypic, it is highly likely to be given a distinct folk generic name.
> 2. The generic name will be restricted in its range of application to the single monotypic genus, x. (pp. 85–86)

The Solega bird data do not support these predictions. We were fortunate to have access to a published, comprehensive and up-to-date list of birds inhabiting the B. R. Hills, which supplied us with reliable information on scientific species names [82]. From this list, we were able to determine which species of birds belonged to locally monotypic genera. Examining the named bird species shown in Table 4.1, it is apparent that around half (46 %) of the *genera* which had Solega bird names were monotypic (Fig. 4.2). It is noteworthy that a similar analysis of the unnamed species from Table 4.5 reveals the same pattern: once again, around half of the genera from this list are monotypic. The implications of this analysis are clear—monotypic bird genera of the B. R. Hills have a roughly 50 % chance of being named in Solega (or not named). In other words, being monotypic has no effect on the likelihood that a particular genus will be named.

Berlin's prediction on the primacy of monotypic genera is further weakened if we count the total numbers of *species* that are named or not named by the Solega. Among the named species shown in Table 4.1, only a third (32 %) belong to monotypic genera, while the majority belong to polytypic genera (Fig. 4.2). Among the unnamed species of Table 4.5, 43 % belong to monotypic genera.

Table 4.5 Common, visually striking or otherwise perceptually salient bird species occurring in BRT, that do not have Solega names

English name	Scientific name	Perceptually salient features	Possible perceptual reasons for lack of name
Blue-bearded Bee-eater*	*Nyctyornis athertonii#*	Large, widespread, distinctive call	
Banded Bay Cuckoo*	*Cacomantis sonneratii*	Distinctive call	
Grey-bellied Cuckoo*	*Cacomantis passerinus*	Distinctive call	
Blue-faced Malkoha*	*Phaenicophaeus viridirostris*	Large	Silent
Ashy Woodswallow	*Artamus fuscus#*	Common and often seen perched in small groups on dead trees	Does not make nests out of mud, like other swifts and swallows
Yellow-footed Green pigeon	*Treron phoenicoptera*	Common, seen in flocks	
Fairy Bluebird	*Irena puella#*	Medium-sized, distinctive call	
Gold-fronted Leafbird	*Chloropsis aurifrons*	Common, loud caller	Camouflaged by its green plumage
Bay-backed Shrike	*Lanius vittatus*	Medium-sized, prominent markings	Found only in scrub forests
Eurasian Golden Oriole	*Oriolus oriolus*	Large and bright yellow, prominent calls	
Black-naped Oriole	*Oriolus chinensis*	Large and bright yellow, prominent calls	Uncommon (a winter visitor)
Black-hooded Oriole	*Oriolus xanthornus*	Large and bright yellow, prominent calls; very common	
Black-headed Cuckooshrike	*Coracina melanoptera*	Medium-sized	Found only in scrub forests
Bar-winged Flycatcher Shrike	*Hemipus picatus#*	Distinctive markings	Small
Black-naped Monarch	*Hypothymis azurea#*	Bright blue, very common	Small
Tickell's Blue Flycatcher	*Cyornis tickelliae*	Distinctive colouration, common	Small
Grey-headed Canary Flycatcher	*Culicicapa ceylonensis#*	Bright yellow, common	Small
Nilgiri Flycatcher	*Eumyias albicaudatus*	Blue	Restricted to high-altitude forests
Common Iora	*Aegithina tiphia#*	Bright colours, prominent calls	Small

(continued)

Table 4.5 (continued)

English name	Scientific name	Perceptually salient features	Possible perceptual reasons for lack of name
Indian Blue Robin	*Luscinia brunnea#*	Distinctive colouration, loud, distinctive call	Winter migrant, rarely seen
Yellow-browed Bulbul	*Iole indica#*	Bright yellow belly, common in moist and evergreen forests, move in loud flocks	
White-browed Bulbul	*Pycnonotus luteolus*	Medium-sized, distinctive call	Hard to spot, usually only in dry forests
Black Bulbul	*Hypsipetes leucocephalus#*	Common in evergreen forests, large size, red beak, crest on head, large noisy flocks	
Grey-headed Bulbul*	*Pycnonotus priocephalus*	Medium sized, distinctive call, flocks in moist forest	Hard to spot, moves within thickets
Jungle Prinia	*Prinia sylvatica*	Common	Small
Grey-breasted Prinia*	*Prinia hodgsonii*	Common	Small
Oriental White-eye	*Zosterops palpebrosus*	Common, distinctive markings, move in flocks	Small
Brown-cheeked Fulvetta	*Alcippe poioicephala#*	Common in moist and evergreen forests, distinctive call	Small
Madras Bush-lark	*Mirafra affinis*	Medium sized, and conspicuous displays in the breeding season; elaborate calls	Not brightly coloured; uncommon
Red-Winged Bush-lark	*Mirafra erythroptera*	Medium sized, and conspicuous displays in the breeding season; elaborate calls	Not brightly coloured
Ashy-crowned Sparrow-lark	*Eremopterix grisea#*	Medium sized, and conspicuous displays in the breeding season; elaborate calls	Not brightly coloured
Rufous-tailed Lark	*Ammomanes phoenicurus#*	Medium sized, and conspicuous displays in the breeding season; elaborate calls	Not brightly coloured
Malabar Lark	*Galerida malabarica#*	Medium sized, and conspicuous displays in the breeding season; elaborate calls	Not brightly coloured; uncommon

(continued)

Table 4.5 (continued)

English name	Scientific name	Perceptually salient features	Possible perceptual reasons for lack of name
Olive-backed Pipit	*Anthus hodgsoni*	Medium sized, common	Well camouflaged
Tree Pipit	*A. trivialis*	Medium sized, common	Well camouflaged
Taddyfield Pipit	*A. rufulus*	Medium sized, common	Well camouflaged
Long-billed Pipit	*A. similis*	Medium sized, common	
White-rumped Shama	*Copsychus malabaricus*	Common, medium sized, distinctive colouration, loud, distinctive call	Not easily seen
Eurasian Blackbird	*Turdus merula#*	Medium sized, common in evergreen and moist forest, glossy black plumage	

Birds not included in the picture elicitation task are marked with an asterisk; # after the scientific name indicates a locally monotypic genus. The scientific identifications are from Grimmett and Inskipp [168]. Any bird up to the size of a sparrow is characterized as 'small'. Birds around the size of a myna are called 'medium', while larger birds are called 'large'

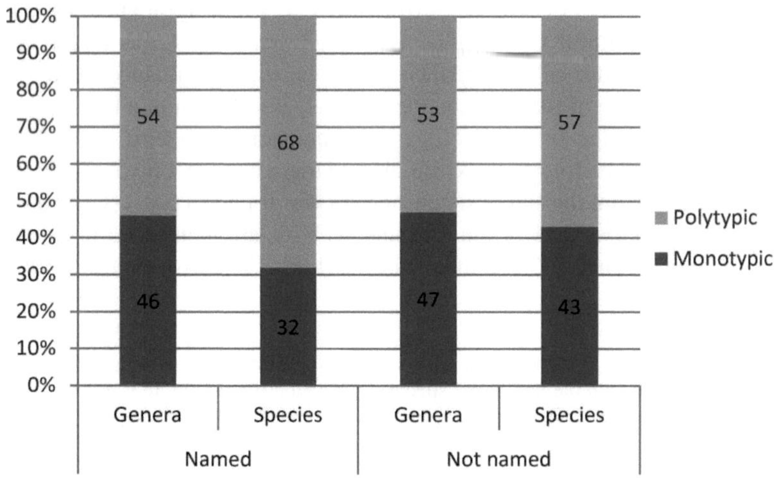

Fig. 4.2 Distribution (in %) of monotypic and polytypic genera among the named and unnamed birds of Tables 4.1 and 4.5 respectively. The *columns labelled 'genera'* indicate counts of genera from the two groups; the *columns labelled 'species'* show counts based on total species number. For instance, *Sitta frontalis* and *Sitta castanea* from Table 4.1 would be counted as 1 genus (for the first column, 'Named—Genera'), but as two species (for the second column, 'Named—Species') under the category 'polytypic'. Previously published in [167]

Berlin's second prediction on the denotational range of generic labels, as applied to monotypic taxa, is also challenged by our data. There are many instances in Table 4.1 where a Solega 'generic' label for a monotypic genus can also be used to refer to other species, which may or may not belong to other monotypic genera. Examples include *maṇṇakki*, which has among its referents two monotypic genera, and two species from polytypic genera; *gumma*, which is the shared name for three monotypic owl genera, and *saṇeyã*, the shared name for one monotypic bee-eater genus and two species from another genus.

4.5 The Effect of Culture

Interesting patterns emerge when we examine the species which participants in the picture elicitation task identified with high or low levels of consistency (see Methods section for a working definition of 'consistency'. For the purpose of the current analysis, species which elicited the same (or similar) labels from three or more villages were deemed 'consistent', while those with fewer than three similar responses were classed as 'inconsistent' (Table 4.6).

Table 4.6 shows that 53 out of 100 biological species were consistently identified by the residents of three or more villages. In all, 35 Solega names were elicited for these 53 bird species. As the participants were asked to volunteer any items of folklore associated with each bird name, we were able to assess the relative cultural importance of species that are readily identified, in contrast to species that are difficult to identify. A total of 20 species from the 'consistent' group had some form of folklore or other cultural significance associated with them—this included stories that explained a bird's call, avoidance practices, the significance attributed to a bird's call, invocation of the bird's name in ritual, association of a bird with a deity, and so on. Knowledge of the ecology and behaviour of a bird species was not included under the heading of 'folklore'. Using these criteria, only one species—the Brahminy Kite—from the 'inconsistent' group was found to possess any associated folklore. In summary, 42 % of the 'consistent' birds are mentioned in Solega folklore, compared to only 2 % of 'inconsistent' birds. The disparity is even greater

Table 4.6 Bird species identified with high and low consistency by participants in the picture elicitation task, along with the number of species with which some folklore is associated

	Total count	Species with associated folklore	%
Consistent species (3+ villages)	53	22	42
Inconsistent species (<3 villages)	47	1	2
Consistent folk names (3+ villages)	35	18	51
Inconsistent folk names (<3 villages)	25	1	4

The total of 60 Solega folk names is obtained by adding the 'generic' names from Tables 4.1 and 4.2

when the Solega names are used as a base count rather than the scientific names—18 out of 35 'consistent' Solega names (51 %) find a place in Solega folklore, compared to only 1 of 25 'inconsistent' names (4 %). We conclude that birds are more likely to be recognised and identified consistently across Solega communities if they have some cultural significance.

These results support Bulmer's claim [173] that salience is a result of many cross-cutting variables. Some of these are purely perceptual (e.g. large size, bright colours), some have ecological and behavioural considerations (e.g. contrast between terrestrial and arboreal, or herbivores and scavengers), and yet others are anthropocentric (e.g. being involved in interactions with humans, or being a source of products desired by humans). Even if it were possible to accurate isolate the perceptual variables from all others, there might still be a considerable degree of overlap between the two groups of variables, at least for some cultures. For instance, the largest birds might be the ones that provide the most meat, while the brightly-coloured species might be highly prized for their feathers. In such cases, it would be pointless to try and tease apart the two dimensions of 'culture' and 'perception', as they would be inextricably linked.

4.6 Relation to Scientific Taxonomy

It is often stated that folk and scientific classification systems show a remarkable degree of convergence at the level of the folk generic taxon; in recent years, such a view has come to be taken for granted, and authors now rarely make the effort to point out where these convergences might lie. Looking at Table 4.1, it could be argued that Solega bird classification closely matches Linnaean classification, on the basis that 45 out of 67 (67 %) Solega bird names (counting both 'generic' and 'subgeneric' names) match a single scientific species. This, however, would be a trivial conclusion, as it merely indicates that something that has a name in one system also has a name in another system. Considerations of space prevent us from entering into a discussion on whether or not Linnaean species constitute 'real' units of nature, although Bulmer [174] provides a well-argued account of why folk taxa are just as real as biological species.

It is important to remember, however, that ethnoclassifications reported in the literature are not the result of omniscient researcher/consultant partnerships casting their gaze on entire species and ecosystems at a time; rather, the subjects of scrutiny tend to be individual organisms (birds, in this case), either living, dead, or photographed. It would be an extremely difficult task to empirically and reliably determine the referential limits of all the folk taxa in an ethnoclassification, just as it is often a very challenging task to ascertain where one biological species begins, and another ends. A more robust test would be to compare larger groupings of 'generic' taxa, as was done for plants in Chapter 3, to see if these also match with the same level of congruence. Using this criterion, it appears that there are some similarities to be found between the Solega and Latin classification schemes presented in Table 4.1. Prominent examples include the drongos (*karaḷiyakki*, genus *Dicrurus*), the barbets (*kuṭrakki*, genus *Megalaima*), the mynas (*uṇṇigirikanakki*, genus *Acridotheres*) and the nightjars (*kapuṭa*, genus *Caprimulgus*), all of which show congruence between the 'folk generic' category and the scientific genus. However, such a correspondence needs to be interpreted with care, as a scientific taxonomy, rather than being an objective 'etic grid' of facts, is more a set of working hypotheses that is continually updated and revised depending on prevailing evolutionary theories and the availability of new data [142].

Making reference to the taxonomy of Indian birds, Ghorpade [175] has warned that it would be a grave mistake to use "*current species names ... promiscuously and uncritically ... as correct, precise*" (p. 982). Most current field guides use outdated taxonomies, with the result that the Indian "*avifauna could number anything between 1500 and 1700 or more confirmed species (including those in Burma and Afghanistan, which belong to the Indian subregion), and not just 1200–1300, as is presently understood and recorded in recent commercially oriented 'DIY' guidebooks*" (p. 982). Although Ghorpade's paper deals specifically with the phenomenon of polytypic species (i.e. true species that have been demoted to the level of subspecies or race), his comments also apply to the "*poorly analysed*" bird genera of India, many of which were excessively lumped together in the first half of the twentieth century, thereby reducing the total number from 600 to around 400.

A good example of this tendency to lump taxa together can be found in the taxonomy of the sunbirds, most of which have traditionally been placed in the large genus *Nectarinia*. Using this label, it would be easy to conclude that there is a good match between the Solega label *tu:gusiṭṭe* and the scientific genus, of which there are four species present in BRT (Table 4.1). The genus *Nectarinia* has now been revised, and many of its members have been assigned to other, new genera. Consequently, the four species recognised by the Solega as *tu:gusiṭṭe* now correspond to the two separate genera, namely *Cinnyris* and *Leptocoma*, with a greatly reduced correspondence between scientific and folk categories [176]. Such revisions have occurred for the avifauna of many parts of the world (e.g. [177] for Mexican birds), but the birds of Asia in general have received little recent taxonomic attention. It is not possible for the average ethnobiologist or linguistic fieldworker to predict which scientific taxa might undergo revision in the future, and it is therefore highly probable that many taxa which currently match folk categories will be revised in the foreseeable future.

Finally, it is worth pointing out that many named Solega categories in fact show a poor correspondence with scientific genera, as they are currently understood. These include *maṇṇakki* (four genera of swifts and swallows), *gi:ṇa* (two genera of parrot/parakeet), *koṭã* (three raptor genera), *marakuṭuka, sivanakki, maragosappa:na* (five woodpecker genera), *giḍugã* (two raptor genera), *gumma* (three owl genera) and *siṭṭe* (at least five genera of small passerines). Moreover, eight species of dove/pigeon in four genera are labelled *araḷakki, soreyakki, bu:jore, pa:riva:ḷa* and *kuggilu*, with little correspondence between folk generics and scientific genera, while the common Yellow-Footed Green Pigeon is not named (Fig. 4.3).

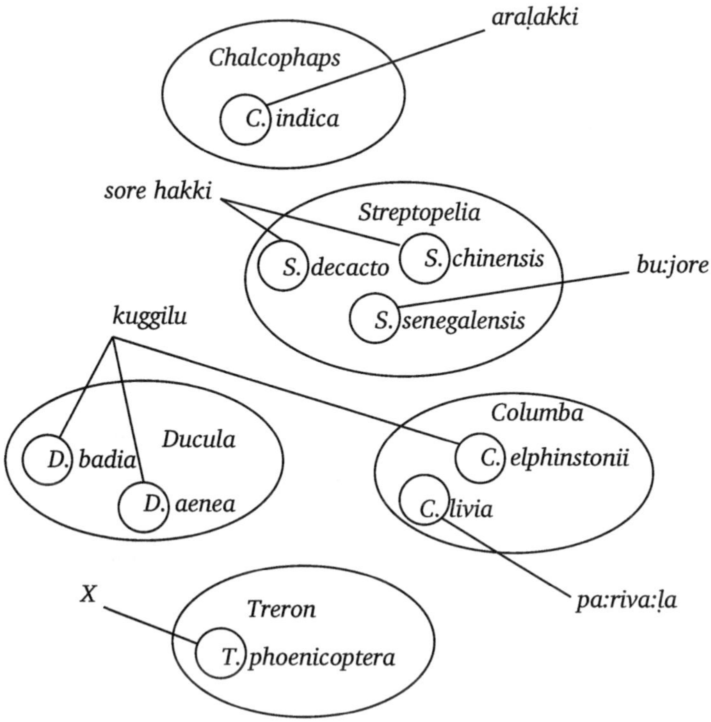

Fig. 4.3 Pigeons/doves found in BRT, and their Solega labels, if any. Previously published in [167]

4.7 Birds in Solega Life, Myth and Ritual

In the final section of this chapter, we describe the very important status that birds enjoy in the secular and religious life of Solega speakers. Birds, in general, play two important roles in Solega folklore: the first is that of a herald of important natural phenomena, both benign and noxious, while the second is that of a moral arbiter that sounds the alarm when important social or cultural taboos are broken. One might speculate that the reason birds are said to possess such gifts is the position they occupy in the physical environment—perched on trees and other high places, they observe not only the world of humans, but also the activities of other animals. As birds communicate their messages to humans through their various calls, it is not surprising that a good number of their vocalisations have been lexicalised as ideophones in Solega (Table 4.7).

As far as we can tell, there is no special purpose symbolic classification scheme in Solega. During our fieldwork, we did not record any lexemes that labelled categories of 'maleficent birds' (along the lines of the nocturnal birds that according to the Nage make the sound *po* [63]), 'birds that warn of danger' or 'birds associated with a god', for instance. Nor were there any named utilitarian categories such as 'birds

Table 4.7 Bird calls that have been lexicalised in Solega

Bird call	Species
aṭṭ-aṭṭu kere suṭṭ-suṭṭu kere	Red Spurfowl
chu chucca	Oriental Scops Owl
jirr-jirr-jirri	Quail
*ka:kanakoṭe**	Indian Scimitar Babbler
karra kurra tikada meṇa ki:ḷu	Large-billed Crow
kẽ kẽ gakku gakku *katale karu kaṭṭu*	Junglefowl
sankara sara tappu	Rooster
ke:ta sattu makka keṭṭu	Indian Cuckoo
ke:tike ku:sa:ta	Eurasian Collared Dove
kisi kisi	Blue-winged Parakeet
kiṭri:/piṭri:/siṭri:	Flameback woodpeckers
*koṭṭipiḍiyã** *also koḍtini - koḍu*	Red-whiskered Bulbul
kuĩ kuĩ	Rufous Woodpecker
*kuruḷi kuruḷi**	Quail
*kurubigo:ḍi**	Rufous Treepie
*kuṭru kuṭru**	Barbets
koho-koho *ḍak-ḍak*	*ko:kunji* (unidentified)
giji giji	Jungle Babbler
*guḍuga:ḍe**	Type of quail/francolin?
*ma:vã kena/pa:pira**	Common Hawk-cuckoo
*pikku pikku**	Wagtails
sirpuru sirpuru	Jungle Babbler
*nanna ku:so**	Puff-throated Babbler
soĩ soĩ/si:ḷ	Parrots, Whistling Thrush

Asterisks indicate bird calls that also provide names for those birds

that should not be eaten' or 'birds that tell the time'. Note that such categories do infrequently occur in the classification of plants and fungi, as all inedible mushrooms are called *uccaṇabe*, while plants with edible greens are often referred to as *soppu*. In the following sections, we outline some well-known pieces of Solega folklore related to birds, which had such a great bearing on whether or not a particular bird was recognised in our picture task. These stories and cultural beliefs were documented in an opportunistic manner, and at present we lack the data to determine whether any significant variation (comparable to the variation in nomenclature discussed above) exists in this domain of folklore. However, anecdotal evidence—whenever the same story was recorded from two or more individuals—suggests that apart from the usual inter-individual stylistic variation that can be expected of the storytelling speech genre, there is a general consensus regarding the content and intent of the stories.

4.7.1 Birds as Sources of Information

Some birds are regarded as important sources of useful information in Solega folklore. The five woodpeckers found in the B. R. Hills may all be labelled *marakuṭuka* in Solega, but two species are given special names that reflect their cultural significance (Table 4.5). Note that Table 4.5 shows a significant degree of ambiguity relating to the identity of these birds—the call of the birds, rather than its appearance, seemed to be the more important identifying feature, but there was still much debate on the subject of which piece of folklore belonged to which bird species. One bird, possibly a kind of Flameback Woodpecker, is also referred to as *ka:ri hakki* 'calling bird', as it is said to be able to predict or induce the onset of rain through its call. Another species of Flameback Woodpecker is called *sa:vakki* 'death bird', as its calls are said to signal the death of a person. Some Solega will also say that the *ka:ri hakki* is able to warn humans of the presence of danger in the form of wild animals. For this reason, it is also sometimes referred to as *sivana hakki* 'Shiva's bird'. Although there was a surprising amount of confusion regarding the precise identities of these birds, two points that all participants did agree on was that there was one woodpecker that called the rain and one that heralded death, and that these birds had very different calls. The Rufous Woodpecker, which looks very different from the flamebacks, was also sometimes given the label *ka:ri hakki*. This is not the only bird to which this power is attributed, as barbets are also said to protect humans from harm (see below).

> *E:vondu kaḍeli biḍu biddo:dare "kuĩ kuĩ" entadu. Adara sama:sa:ra to:rtera:du. Maḷe a:gottu, a:ga kirri-enna karetade, no:ḍi, a:va:ga maḷe de:sti (ka:ri hakki)*
> When it gets overcast somewhere, it says "*kui kui*". It relays the news. When it rains, then it calls "*kirri*", see? Then it rains a lot.

> *Hakki piṭri:-ennade. O:, de:lo? Hakki siḍdadella? E:ka? I: da:ri ho:ku a:ga hakki siḍdattu. O:, alli ho:ga ba:radu, de:nondu ade alli. Alli ho:da:giddale a:neve yo, e:nondu karaḍiye yo, e:nondu ja:ti kaṇḍe uḍtade a:ga. Matte i: da:ri ho:ga be:ku endu yo:cane ma:ḍida:ga e:na ma:ḍade, "kirri-kirri-kirri" endade. O:, i: da:rili e:nu ille. (ka:ri hakki)*
> The bird says "*piṭri:*". "Oh! What's that? The bird's calling? Why?" You want to go along a path and the bird calls (and so you think), "Oh! I shouldn't go there, there's something there." If you keep going you encounter some wild animal, whether it be an elephant or a bear. And when you want to go along a path and the bird says, "*kirri-kirri-kirri*", (you say to yourself) "Oh, there's nothing here."

> *E:va kaḍeli maraṇa a:godare, sa:vu a:dare, adu suddi koḍtade. ade emadamana ko:ḷi ennadu (sa:vakki)*
> When there's a death somewhere, it gives a signal. That's why it's called the bird of the god of death.

Here, it is worth noting that the actual call of the rufous woodpecker is best characterised as "*kuĩ kuĩ*", and that of the greater flameback as "*kiṭri:*".

Barbets are important in Solega folklore, not only as protectors of humans against wild beasts, but also as sources of practical information. In the first of the following two extracts, the barbet (in particular, the White-cheeked Barbet) is described as an accurate timekeeper, with its late afternoon and evening songs telling humans that it is time to stop work and return home from their fields. Three separate bouts of singing are said to emanate from the barbet's throat at this time of day, and although the first two bouts are given unique names, the speaker was unable to explain their significance. In the second extract, the speaker explains that the barbet's song, much like that of the Flameback Woodpecker, can help humans avoid running into dangerous wild animals.

Sande a:da:ga a:ga idu mo:ḍa ka:vaḷa allava? A:ga kuṭru ha:ḍdade. "O:! I:ga aḷiyãna kuṭru ha:ḍdade! I:ga time *a:gottu kaṇo!" A:ga puna innondu suma:r ottu iḷdu, a:gondu kuṭru ha:ḍdade. Idu ma:vana kuṭru. Kuṭrina saddakave* time *a:gottu endu thiḷdumaku na:vu … a:seka eraḍane kuṭriga puna mu:rne kuṭriga* sound *bandale a:ga kattale a:gottu. Mu:re* sound *koṭṭade adu … a mu:rne* sound *biddottu enda:ga, aḍige teya:r ma:ḍa:ku. (kuṭrakki)*
It gets dark in the evening, right? That's when the barbet sings. "Oh! That's the son-in-law barbet singing now! It's time!" Then, when the sun sets completely, another barbet will sing. This is the father-in-law barbet. We tell the time just from listening to the barbet's song … when the second barbet sings, and then the third barbet sings, it gets dark. It only sings three times … when you hear that third sound, you should start cooking.

Adu suddi koṭṭre a:vottina dina yelli iddarue huli kirubã aḍḍa baḷsi odde (kuṭrakki)
When the barbet calls, no matter where you are that day, tigers and leopards will avoid you.

Another timekeeping bird is *ke:sanakki* 'Indian Cuckoo'. Unlike the barbet, however, whose call signals a diurnal rhythm, the cuckoo's song waxes and wanes with an annual period. This bird is heard only in the months of April and May, and the presence of its song heralds the first rains of the year, or *munga:ru*. The content of the four-syllable song, on the other hand, which is rendered in Solega as "*ke:ta satto, makka keṭṭo*" has a more sinister meaning.

Ke:ta satto, makka keṭṭo. Tande satta me:le avana makka bi:de pa:lu anta. (ke:sanakki)
Ke:ta died, his children cried [lit. things were bad for his children]. After the father died, the children were left destitute.

We were told of a bird called *ko:kunji* (which we have been unable to identify, but is possibly the Slaty-legged Crake) in one village, which was said to have the ability to forecast the rain, just like the woodpecker. The 'rain' vocalisations are said to be heard at night; the vocalisations heard during the day indicate the absence of rain:

Ko:kunji ante:ḷidale ondu sa:istra ru:padalli—bu:mi me:le ku:tkoṇḍu koho-koho-koho enna ku:gidare, maḷe ja:sti ante:ḷi, adu ha:ra:ḍkoṇḍu ku:gidare, maḷe kaḍame.
Ko:kunji can make forecasts—if it sits on the ground, and calls *koho-koho-koho*, it means there will be plenty of rain, but if it calls while flying, there won't be much.

4.7.2 Birds with Supernatural or Ritual Connections

Large owls, *gumma*, are also birds of ill omen, possibly due to their human-like faces, which are usually seen at night. Solega people may avoid looking at even a picture of a *gumma*, and the threat of this bird is often used to frighten young children into obedience: "*gumma bandade, summane malagu*" (The *gumma*'s coming;

go to sleep!). The Solega, particularly those who live on the Tamil Nadu side of the border, say that the call of a *gumma* signifies the presence of *e:ḷu kula* 'Seven Clan' Solega (also called *uraḷi solegavaru* or *kurubaru*) in the forest. The Solega living in the B. R. Hills are called the *aidu kula* 'Five Clan' Solega, and speak a language that is closely related to the official state language, Kannada. They regard themselves as loosely affiliated with, but distinct from, the *e:ḷu kula* 'Seven Clan' Solega, who mostly live in Tamil Nadu, and whose language has much in common with Tamil. These people are thought to be skilled in black magic and traditional medicine, and it is said that they can render themselves invisible. The Five-Clan Solega are afraid of these people, and believe that their men would kidnap young girls for marriage, and steal the grain from their fields. So if one hears a *gumma* calling while watching over one's crops at night, one dares not shoo away any animal that fed on that crop that night, as it could be one of the *kurubaru* in animal form. A *gumma* calling a single time signifies the presence of the *kurubaru*, but if it calls twice, it is just a regular *gumma* that means no harm. Negative attitudes directed towards owls have also been reported from other parts of the world—an owl is regarded as the watchman of a witch or sorcerer by the Mbuti of the Congo basin [178], while the Tembo, who also live nearby, claim that "sorcerers use it [the spotted eagle-owl] as a telephone" [179]. The Sahaptin of the Pacific Northwest consider owls to be "a dangerous being, an Indian doctor and an omen of death" [180], and Forth [63, 69] similarly reports that the Nage of Indonesia speak of owls as the form assumed by the maleficent spirit of a witch. Interestingly, the Nage also attribute special significance to the number of times the call of an owl is heard, as an uneven number of calls is meant to indicate the presence of a witch nearby. In many parts of India, however, equating someone with an owl simply means that s/he is stupid.

Mystical powers are attributed to other birds such as the fish and hawk owls *gu:be*, (possibly) the Large Cuckooshrike, referred to as *kaṇiga:rã* and the *ko:ṭã*, or Oriental Honey-buzzard. The *ko:ṭã*'s unusual night-time courtship vocalisations, consisting of a long series of rapid, staccato bursts whose pitch varies over time, are said to be a sign that the bird has been possessed by a god.

Ondu ka:la ettadu ondu ka:lu hi:ge. kuṇiyadu, batta kuṭṭadante adu, de:va gaṇatige. (gu:be)
It has one leg raised up, and one leg (down) like this. It dances, it grinds rice flour, as an offering to the gods.

I:ga na:vu beleka tekobe:ka:giddale, a:ga saistra no:ḍista:re. avã saistraga:rã. (kaṇiga:rã)
When we want to take our produce (to the market), it tells us how we will do. It's the fortune-teller.

Ko:ṭã ku:gidale ko:ṭi ba:ḷa:ṭa (ko:ṭã)
When the Honey-Buzzard calls, you get ten million blessings.

The Emerald Dove is another bird with divine connections. The Solega name for this bird, *araḷakki* or 'castor bird', is a reference to its habit of feeding on the seeds of the castor plant, which grows wild, but is also cultivated by the Solega. The following myth explains that the bird collects castor seeds not for its own sustenance, but for a higher purpose. Participants from one village commented that this bird can fly at great speeds, but also warned against trying to imitate the bird's call.

Avã araḷakki endale araḷaiyã. Avã sva:mine, avana pu:je ma:ḍta:re. Avã araḷaiyyãnatte avã srushṭi ma:ḍidã. I: araḷa a:du a:du a:du se:rsi koḍadu. Eṇṇe illa:doitu avana maṇḍeka. Araḷina eṇṇe be:ka:gittu, adakkatte araḷakki kaḷisinã. I: hakkiya kaḷisi puna i:kaḍe a:kaḍe buṭṭu, puna avã tandade akki ella, araḷella adu tandu me:sadalli alli etti surdattu keḷage avanagaḍe.

The castor bird is associated with Grandfather Castor. He's a god, and we worship him. It was Grandfather Castor who created it. It painstakingly collects castor seeds, and gives them to him. He didn't have oil to put in his hair. He wanted castor oil, and that's why he sent the castor bird. He sent the bird here and there, and the bird brought him all the castor seeds in its crop, and poured them out in front of him.

Aṇakisadaka ue a:ga:dille. Kivĩli u:ḷe surdu uṭṭade endu. (araḷakki)

You can't repeat what it says. It is said that it makes earwax start to flow out of your ears.

A large and common bird, the Greater Coucal, or *kembutã*, can also bring about ill luck, due to its association with the death of an ogre. The ill luck can, fortunately, be countered by means of a simple ritual.

Savaṇana kola:ga sva:mi iddã. No:ḍidã, balagaḍinda oṇṭottutte atta:gi, hinde savaṇanige e:ṭade. Balagaḍinda oṇṭo:da:ga na:vu aga:darue, ondu kone murdu—muḷḷugone ue sarie, ondu kalla:dalue sarie—adava etti attaka tiṭṭu uḍa:du. Nammadu da:ṇṭa:gille. Adakka e:na ennabe:ku endale, "Ni: ho:da da:ri kallu muḷḷu a:guḍa be:ku, na:nu ho:da da:ri sa:lu sampage vana huṭṭi santo:sava:gi beḷiya be:ku," enduru buḍa be:ku. A: tara ma:ḍadiddale e:na:daru tondari ade. Adu ondu karmada hakki.

When (the ogre) *savaṇa* died, the Lord was there. He saw a coucal passing by his right-hand side, when *savaṇa* lay dying. So when we see a coucal on our right, we break a branch—it can be a thorn or even a stone—and place it there (on the ground). Then it doesn't cross our path. What that means is, "May your path be full of rocks and thorns, and may a forest of *sampage* saplings grow blissfully along mine." If you don't do that, there will be trouble ahead. It is a fateful bird.

A number of prominent Solega folksongs, which are sung at festivals and celebrations, make reference to birds. The simplest of these songs honours the mud-nest-building swifts and swallows, or *maṇṇakki*, that are common in the B. R. Hills. For reasons unknown, the birds are referred to as *maṇṇakki ma:vã* 'father-in-law mud-bird' in the song, which is open-ended, and can be repeated any number of times by naming different tree species in the second line:

Maṇṇa tege tege maṇṇakki ma:vã,
Araḷu kaḍḍella maṇṇakki gu:ḍu.
Collect mud, o father-in-law swift,
Swift nests among the castor branches.

A far more important and lengthy song cycle called *ha:ḍuke* is sung at the shrine of a god on his festival day from dusk till the following dawn. The sequence of themes covered in the *ha:ḍuke* is such that the final few stanzas, which consist of invocations of certain birds, coincide with the first calls of those birds in the forest as the sun rises. A recent addition into Solega folklore may be the *garuḍã* 'Brahminy Kite'. This bird plays a key role in an important Hindu (i.e. belonging to non-Solega, plains people) festival that takes place in the B. R. Hills. The ceremonial chariot that is drawn in procession during this festival may only leave the temple after this bird is seen circling overhead. Given that Solega people participate in this festival, but do not claim it as their own, it is perhaps significant that the only village that correctly named this bird as *garuḍã* in the picture elicitation task was the one situated in the lowlands, at the edges of traditional Solega territory.

4.7.3 Birds as Moral Arbiters

The call of the Puff-throated Babbler (called *ku:sakki* by some Solega participants) is also a death lament. It is said that a mother and her baby, both exceedingly thirsty, stopped to rest, and that the mother, in her haste, proceeded to drink some water without first seeing to her infant. Her act of selfishness caused the infant to die. Stricken with grief at her failure to care for her own baby, the mother turned into a bird, and to this day, calls out *nanna ku:so nanna ku:so* 'My child! My child!'

> *Obba awwe maga tabburu ho:da. Ni:rutte da:va hattira* time-*inalli ni:r etturo:gi e:ri mele nilsu uṭṭu, avatte ho:gi rappenna kuḍiya:ge biddu uṭṭa. Maguvina usura oṭṭottu. Ni:ra tãkuḍdu uṭṭu itta:ka no:ḍida, ku:sina usura oṭṭottu. Adakka etti* first-*e: kuḍsuṭṭiddale seriya:gittu, ade:nue tondari a:galle. Ta:nu kuḍdu uṭṭa, idakka usuru oṭṭottu. A:ga porr-enna a:ri ho:gi, alli ku:turu,, "Nanna ku:so:! Nanna ku:so:!" ennadu ha:ḍa:du.*
> A mother was walking along, holding her child in her arms. In that time of great thirst she rested on a hillock, and there she eagerly drank the water that she had brought along with her. (At that moment) the child died. By the time she had finished quenching her thirst, and turned to look at her child, it was already dead. If she had given her child some water to drink first, everything would have been fine. But she drank first, and that's why the child died. Then she flew up, and sitting there (on a branch), started to sing, "My child! My child!"

Two other birds serve as witnesses to human frailty in Solega folklore—these are the Common Hawk-cuckoo *pa:pira* and the Large-billed Crow *ka:ge.*

> *Ma:na mariya:di ma:tu idu. Ma:vãnuve soseuve ibbaruve elli ho:ga* time-*inalli, a:ga sose ondu kallu me:le ku:tu keḷeka negadu uṭṭa. "Husha:ravva!" enda:ki hindeka turugi no:ḍida:ga avaḷa ma:na kaṇḍu uṭṭattu. A:seka ma:vãniga, "O ho!" iḍiya:gave avka:sha a:goitu. Soseya gabbakenna iḍiya:da:ka:ue, ondu hakki, alliye: ku:tidu, "Ma:vã ke:na! Ma:vã ke:na! Ma:vã ke:na!" endale "pa:pira: pa:pira: pa:pira:" endadu. Sa:kshi koḍadadu, a: hakki. Puna hoṭṭogiga ibbaru soseue ma:vanue meneka bandu, sose ma:ŋana mene hoṭṭoda, matte ma:vã ma:vãna meneka hoṭṭonã. A: hakki hullina su:ru me:le ku:taru, "Ma:vã ke:na! Ma:vã ke:na! Ma:vã ke:na!" ha:ḍdade, endale ma:vã soseka oṇṭogu uṭṭã. (pa:pira)*
> This is a tale of morals and propriety. A man and his daughter-in-law were travelling; after resting on a rock, the girl proceeded to jump down. "Careful!" said the man, and as he turned towards her, caught a glimpse of her breast. Sensing his opportunity, the man thought to himself, "A ha!" As he quickly grabbed her, a bird, sitting above, cried out, "*Ma:vã ke:na! Ma:vã ke:na!*" or "*pa:pira pa:pira pa:pira*". It witnessed the act, that bird. Later the man and his daughter returned home, and the girl went to her husband's house, while the man returned to his own. The bird landed on the grass roof of the house, and sang, "*Ma:vã ke:na! Ma:vã ke:na!*", meaning that the man had had relations with his daughter-in-law.

The crow story starts with a woman who is asked by her husband to cook a meal for some guests. Left on her own, the woman greedily ate most of the beans and *ra:gi* balls while cooking, with the result that there was very little left for the guests. Later that night, she suffered from severe indigestion, and was unable to sleep. She was afflicted with a debilitating amount of flatulence, which she tried to remedy by blocking her anus with a plug made from the wax of *nesari je:nu*, the stingless bee. Unfortunately, enteric gas built up inside her belly, and the next morning, she was found dead by her husband. As people prepared to bury her, a crow, who had been watching the proceedings, called out, "*Karra kurra tikada meṇa ki:ḷu*" "*Karra kurra!* Remove the wax from (her) bottom!". An old woman came along, removed the plug with a stick, and cast all the escaping gas onto a nearby *si:ge* bush. She then dropped dead, but the woman who had eaten the beans sat bolt upright, and was well again. Incidentally, the leaves of the *si:ge* bush took on a foul odour, and for that reason is now called *ũsu si:ge* 'flatulence *si:ge*'.

4.7.4 Miscellaneous Folklore

Some birds are afforded a privileged position by the Solega on the basis of observed behavioural and ecological traits. The Greater Racket-tailed Drongo, *doḍḍa karaḷi*, which is able to mimic the calls of other birds with great accuracy, is thought to use this ability to attract birds of other species into mixed-species flocks [181]. The significance of such unusual gatherings of birds, and of the possible role of the drongo in maintaining them, is not lost on the Solega; the drongo is also called *ko:luka:rã* (lit. 'rod bearer') the title given to a traditional Solega elder charged with maintaining peace and order, and meting out punishment to wrongdoers.

> *Ko:luka:rã anta. Pakshigaḷige ella idu va:ra ettavã (doḍḍa karaḷi)*
> We call it the sheriff. It's like a counsellor to all the birds.

An identical belief exists among the Mbuti of the Congo basin, and for much the same reasons, but for a different bird species that also leads mixed-species groups. According to Ichikawa [178]:

> A species of greenbuls called *mbilie* (*Criniger calurus*) is said to be the chief of the birds, because, according to the Mbuti, other birds will gather around him, when it calls "mbilie!" (p. 112)

An unrelated point of cultural significance is that the ashes obtained from burning the feathers of the drongos are smeared on the forehead of Solega men who are about to set out on a honey collecting expedition. This may be related to the drongo's habit of flying after migrating bee swarms, and picking off stragglers.

Another bird, the Malabar Whistling Thrush, or *ka:nagoravã*, sings at dawn, and is said to be responsible for leading cows and buffalos out to pasture. Its slow, halting call has a strikingly human-like quality, and has earned it the nickname 'whistling schoolboy' among ornithologists. The Solega, however, say that it resembles the speech of a stutterer.

Goravã endale todavã enta ... Golla, gollarakki adu. Dana, emme, adella idu buḍtade no:ḍi. Beḷagu aidu gaṇṭeka soĩ: enta, si:ḷ-enta buḍta:de. Ba:gilu tegedu ... doḍḍi. (ka:nagoravã)
We call the Whistling Thrush a stutterer… It's the cowherd, a cowherding bird. Cows, buffalos, it leads them all, see? It calls out "*soĩ:*" at five in the morning, it leads them out with a "***si:ḷ***". It opens the door … of the cowshed.

Finally, the highly variable calls of the Hill Myna *se:nanakki* have led the Solega to conclude that English speakers must have learnt their language from this bird. The myna's song was characterised by one speaker as "*hĩ: huĩ kĩ: sĩ: poĩ sĩ: poĩ*", a rendition full of falling pitches and alien-sounding diphthongs.

English-*inavariga ma:tu barade, adu tante ma:ta kalisiradu. Korama ja:ti hakki adu. (se:nanakki)*
When the English couldn't speak, it was that bird that taught them language. That bird is a show-off.

Solega people have traditionally eaten both the eggs and the flesh of many species of birds, although such practices have declined sharply since the setting up of a wildlife sanctuary on their lands. Some birds are never eaten, however, and these include the crow, the pheasant coucal and woodpeckers. This taboo is very likely a direct consequence of the powers, both noxious and benign, that the latter two groups are supposed to possess. In the case of the crow and the coucal, the taboo may also be due to their unusual calls, their scavenging habits (at least for the former) and the unpleasant smell and/or taste of their flesh. Much like the coucal, the crow is regarded as an ill-omened bird, as it is the vehicle of *shani*, the personification of Saturn. As a result, children are warned to not even throw stones at crows. One final culinary taboo among the Solega concerns the head and legs of chickens—these should not be eaten by men, as this will cause their bodies to tremble when they attempt to climb trees.

The bird-related Solega folklore described above can be divided into seven broad categories, namely 'interactions of birds with other creatures', 'useful signs and signals', 'ill omens', 'religious connections', 'moral tales', 'secular songs' and 'pest species'. Table 4.8 shows that Solega bird folklore is dominated by two categories, namely the 'useful signs and signals' and the 'religious', which together comprise just over half the items. Half of Solega bird folklore, then, is devoted to items of strong utilitarian or cultural value. One could easily imagine any language having

Table 4.8 The relative proportions of different types of Solega bird folklore

Folklore category	Count	%
Interaction with other creatures	3	7.7
Useful signals	11	28.2
Ill omens	6	15.4
Religious	11	28.2
Moral	3	7.7
Secular song	3	7.7
Pest	2	5.1

an inventory of myths or folk tales that focus on the morphological features of prominent plants or animals in their environment—'How the peacock got its tail', or 'Why the cockatoo is white', for instance. The only such story that we have come across in Solega is the incident of the elephant and the quail mentioned above. Some Solega have mentioned that the slender curved beak of the *moḍemoka* 'Hoopoe' is indicative of how this bird originated from members of the *cama:r* 'leather tanner' caste, who routinely use similarly shaped needles. Moreover, the items of folklore in the 'useful signals' category have, as their referents, not the appearance of a bird in the B. R. Hills forests (indeed, the annual arrival of the migratory *kokkore* is not mentioned in any Solega folktale), but the vocalisations of certain species.

Evidence from other biological domains supports the view that meaningful signals and interactions—real or potential—are highly salient to the Solega. The calling of a lizard confirms the occurrence of a bad event: if one is talking about the possibility of bad weather, or a bad harvest, and the lizard calls at the same time, then it is a confirmation of that event occurring in the future. The mass chirping of cicadas is a predictor of rain (as in many cultures, the connection between heavy rain and the croaking of frogs is also noted by the Solega), and if a dog is seen barking for no apparent reason, there could be a *daiyya* 'bad spirit' nearby. Here, it is noteworthy that reptiles, amphibians and insects (with the exception of biting ants and the honey-producing bees) do not otherwise figure in Solega folklore. The vocalisations of elephants are also lexicalized as ideophones in Solega, as knowledge of these sounds is invaluable in allowing individuals to avoid these potentially dangerous and unpredictable beasts. The Solega elephant sound inventory includes snorting and trumpeting sounds, as well as the sounds of an elephant flapping its ears, or passing copious amounts of urine.

Bulmer [173] has noted that the bird classification of the Kalam of the New Guinea Highlands is influenced by ritual or mystical significance (e.g. as totems or omens), utilitarian value (as a source of food or feathers) and perceptual salience (large size and bright colours). In addition, Bulmer points out that "*the way in which birds ... interact spontaneously with man, or are interpreted as doing so*" also underlies Kalam discrimination and evaluation of birds (p. 65). Our work with the Solega strongly suggests that they too name and talk about some birds not merely because they are large, common or colourful, but also because they perceive these birds to be a source of meaningful signals or interactions. This may include signals that are received by humans (e.g. in the case of the woodpeckers) or interactions that involve divine beings (e.g. the Emerald Pigeon), other animals (e.g. the quail) or humans (e.g. parrots as agricultural pests). In some instances, the flow of information is bi-directional: the puff-throated babbler, the crow and the hawk-cuckoo are said to signal the breaking of moral codes of conduct by humans after having witnessed the acts themselves. In the case of the owl, the interaction is a potential one, as sighting this bird at night is something to be avoided. In contrast to the broad definition of cultural/utilitarian significance used by Bulmer and by our own study on Solega ethno-ornithology, Berlin [9], trying to downplay the importance of cultural/utilitarian significance, limits his definition to species that are eaten by humans:

> Birds play an even less important role [than mammals, to the Aguaruna and Huambisa of Peru]. Some thirty to forty bird species, mostly the larger game birds, contribute an average of 8 percent by weight to animal consumption, but none of these species contributes more than 3 percent. By far the most dietarily significant bird is the introduced chicken. (p. 89)

This obsession with only 'what is eaten' is worrying, and has been previously noted by Hunn [62] in relation to plants. Solega people also consume more chicken than any other bird, but to argue that this fact alone makes chickens more culturally significant than the woodpeckers, babblers, drongos and barbets that the Solega sing about, listen out for, and tell stories about would be misguided in the extreme.

'Folklore' is probably not an entirely appropriate umbrella term for the diverse items of cultural knowledge contained in the above categories, as many of the Solega beliefs are undoubtedly based on a core of, what a western-trained scientist might consider to be, biological fact. The example of the Greater Racket-tailed Drongo, considered a leader among birds, is a prime example. Some of the useful information that birdcalls are said to relay to humans may also have a biological basis—many species of birds, including the woodpeckers found in the B. R. Hills can produce special alarm vocalisations in addition to their standard calls, which may serve to alert actual or potential mates, or even other members of their flock to the presence of a predator. It has been shown that birds that perceive an alarm call will hide themselves or flee [182], and the Solega's belief that certain woodpeckers' calls can indicate the presence or absence of danger may be based on such observations. Similarly, the frequent co-occurrence of the calls of some birds like the *ka:ri hakki* or the *ke:sakki* with the onset of rain or the start of a new season is frequently attested by visitors to the B. R. Hills. Such an association with rain is even encoded in the formal English names of birds such as the Rain Quail *Coturnix coromandelica*, which appears "*in certain districts coinciding with the opening of the rains... It is generally a bird of the plains, but on the advent of the rains will penetrate up to 6,000 feet in the Himalayas and Nilgiris*" ([183] p. 265).

The relationships between certain birds and plants are sometimes encoded in Solega lexemes, a prominent example being *araḷakki* 'castor bird', whose story was described earlier. Ornithologists have noted that this bird does indeed have a particular affinity for the seeds of the castor plant *Ricinus communis* [184]. The seeds of the castor plant *Ricinus communis* are known to be extremely poisonous to humans and other mammals, but it has been shown that birds have a higher tolerance to this plant's toxin [185]. The regurgitation of castor seeds that the dove performs in the Solega story may indicate a physiological mechanism to deal with the ill effects of ingesting too high a dose of toxin.

There are also instances of plants being named after birds (we suspect this direction of lexical transfer because similar bird names exist in some neighbouring languages such as Kannada, Telegu and Malayalam), although our consultants were unclear about the exact nature of the link between the two species. Examples include *pa:riva:ḷada mara* 'pigeon tree' (*Butea monosperma*) and *gi:jigana giḍa* 'weaver bird plant' (*Caesalpinia mimosoides*), both of which are species commonly found in the B. R. Hills.

4.8 Conclusion

The results presented in the first half of this paper demonstrate that bird naming in Solega is not a straightforward affair, and that it would be naïve to assume that there is a lexicon of universally accepted, 'correct' bird names in the language, simply waiting to be elicited. The initial name list presented in Table 4.1, which was compiled by SA over 5 years of fieldwork with Solega-speaking field assistants, was approved by numerous Solega from villages situated around the field station. The picture elicitation task, carried out in villages located far from the field station, showed that there were other ways of naming many of the birds on our list; the variation we encountered ranged from transparent dialectal differences in pronunciation to completely unrelated lexemes (Table 4.4). Moreover, while we were confident about our field assistants' judgements on the acceptability of omitting—*akki* 'bird' from the generic names shown in Table 4.1, Solega speakers from other villages clearly had different ideas about when this morpheme should be used (Table 4.2). This situation is reminiscent of the ethnobiological knowledge of the Wola people of Papua New Guinea, as described by Sillitoe [60]. Faced with a significant level of disagreement among his consultants regarding the identification and classification of bird species, Sillitoe argues that it is futile to attempt to compile a single ethnotaxonomy for a community like the Wola, which is characterised by intellectual egalitarianism and acephalous politics. In other words, individual Wola are able to decide for themselves what constitutes an appropriate taxonomy, as there is no 'higher authority' to contradict them, and prescribe a 'standard' taxonomy. The Wola display "considerable tolerance of dissonance" (p. 1167), and, as a result, their animal classification scheme:

> is inherently dynamic and subject to negotiation; there can be no closure or final bounded version, no authoritative comprehensive arrangement. (p. 1169)

Many Solega examples discussed above support this view. When discussing the common and culturally-important woodpeckers, Solega speakers seemed quite at ease with changing their minds about which labels to apply to which woodpecker species, and which birds to group together under a single category. The Solega labels *marakuṭuka*, *sa:vakki* and *ka:rihakki* were known to most participants, but there seemed to be only a loose correspondence between the names and the birds. Ultimately, the more important pieces of information were the stories and omens associated with each bird name.

A further example is that of the Paradise Flycatcher (*Terpsiphone paradisi*), which, in our initial name list (Table 4.1) was called *saṭṭugaba:la* 'ladle tail'. To our surprise—as this is a visually highly distinctive bird—this name was not recorded once in any of the five villages visited during the picture elicitation task. Three plausible names were recorded from three villages, while participants from the remaining two villages said that they had definitely seen the bird (and were able to describe its appearance, including its habits and sexual dimorphism), but did not have a name for it. This example also demonstrates how perceptual distinctiveness is not the sole force that shapes folk taxonomies. The same can be said of the brightly-coloured orioles, which are unnamed in spite of being significantly larger than the *ma:dihakki* (Scarlet Minivet). The picture elicitation task showed that a bird was far more likely to be identified and named consistently if it had some sort of folklore associated with it (Table 4.6). Additionally, several prominent or common species of bird present in the B. R. Hills remain unnamed in Solega (Table 4.5), while visually unremarkable birds such as the puff-throated babbler and the quail are prominent in the Solega psyche.

Some other strong predictions made in Berlin [9] are not supported by our data. For instance, belonging to a locally monotypic genus in no way increases a bird's chances of being named in Solega. With regard to nomenclature, we have demonstrated that while bare mononomials dominate the responses in artificial situations such as the picture elicitation task, birds are more likely to be referred to with their full, binomial names in spontaneous speech. Finally, the apparent correspondence between scientific classification and Solega ethnotaxonomy appears to be superficial when larger 'generic' groupings of birds from both systems are compared. While some Solega groupings neatly match their Linnaean counterparts, as they are currently recognised, some show little to no correspondence.

The flycatcher example discussed above contrasts sharply with the case of the puff-throated babbler, which, in spite of its important place in Solega culture, is a 'bird with no name' for many Solega. The bird's call is recognised by every single Solega we spoke with, and the story associated with this bird is also universally known. Yet, many consultants were unperturbed by the fact that they did not know its name; a few provided what appeared to be a nonce coining, along the lines of *ku:sakkl* 'child bird'. Cases such as this illustrate that neither 'perception' nor 'culture' can, in isolation, explain the architecture of such a complex social and linguistic construct as a folk classification. Simple answers to complex problems are always intellectually satisfying, especially when they appear to allow comparisons across disparate systems. Unfortunately, such one-size-fits-all theoretical frameworks also carry with them the risk of making the complex phenomena they seek to explain seem far simpler than they really are. For decades, biomedical researchers vigorously debated whether it was 'nature' or 'nurture' that was responsible for the incredible diversity of physiological, morphological, behavioural and pathological phenotypes that make up the human condition. Both camps were able to cite a great deal of valid empirical evidence to back up their viewpoints [186]; viewpoints that seemed to be so diametrically opposed, as to be irreconcilable. Ultimately, however, all reasonable scientists were forced to admit, when faced with a wide array of experimental studies, that human development is the result of the interdependent action of both heredity and environment [187]. In other words, an organism's genetic makeup bestows upon it certain predispositions, but external factors can radically alter its seemingly pre-ordained developmental trajectory. Recent research in psychology and linguistics has echoed such conciliatory attitudes, with an increasing number of studies demonstrating an effect of culture on cognition as a whole [188], and on language in particular (e.g. [189]). Our data do not lead us to believe that the Solega classification of birds is "*largely immune from the variable cultural determinants found in other areas of human experience*" ([9], p. 9) any more than they support the hypothesis that the taxonomic systems of 'primitive peoples' lack any consistent underlying logical principles [190]. Instead, we acknowledge the different, but interdependent, roles of perception and culture in the creation of folk taxonomies.

Chapter 5
Landscape Terms in Solega

5.1 Introduction

The study of human conceptualisations of landscape, as they are encoded in the world's languages, has the potential to provide answers to some basic linguistic questions, such as:

- Are there common cross-linguistic patterns in the way features of the landscape are perceived and named?
- What aspects of a landscape feature are incorporated into the name of that feature?
- What manner of cultural, utilitarian or other encyclopaedic information is associated with a named landscape feature?
- How are different landscape features related to each other in people's minds, and how do these relationships manifest themselves in language?
- Are there regular patterns in the way landscape features are given proper names, and how do named landscape features, in turn, influence the naming of other entities?

Many other questions have been posed by authors interested in investigating the relationships between people, landscape and language. Three important collections of language-centric studies on landscape have been published in recent years: the first is a special issue on 'Language and Landscape: geographical ontology in cross-linguistic perspective" in the journal *Language Sciences*, while the second is an edited volume called "Landscape in Language: transdisciplinary perspectives". In their introductory article in the *Language Sciences* special issue, Burenhult and Levinson [104] argued that landscape did not appear to be an independent semantic domain like kinship or colour. Nevertheless, the fact that, in many of the languages studied to date, culture, language and utilitarian considerations can all contribute to some extent to the categorisation of landscape features, makes landscape a particularly interesting field of investigation. The introductory chapter of the

A. Si, *The Traditional Ecological Knowledge of the Solega*, Ethnobiology,
DOI 10.1007/978-3-319-24681-9_5

Landscape in Language volume [191] presents the aims of the young field of 'ethnophysiography', and these are said to include not only geographical ontology and the precise semantics of terms used to denote landscape features, but also "*the study of knowledge systems, beliefs and customs of a people concerning landforms and landscapes*' (p. 7). Finally, the recent edited volume *Landscape Ethnoecology* [192] continues this theme by exploring not only people's interactions with various landscapes, but also their knowledge of the ecosystems to be found on different landscape types (see especially [193] for an overview).

In addition to these language-centred questions regarding human perceptions of landscape, there exists a large body of literature that deals with the psychological correlates of humans' interactions with their surroundings. The concept of 'cognitive maps' has been used to describe the "*long-term memory representation of the properties of and the components in the environment*" [194], and in spite of the numerous competing definitions for this concept [64], it remains a useful heuristic with which to probe a significant domain of human language and cognition.

> Kitchin [64], in his review of this field, distilled existing orientations concerning cognitive maps into four main viewpoints:
>
> (1) Is it the case that the cognitive map *is* a cartographic map (Explicit statement)?
> (2) Is it the case that the cognitive map *is like a* cartographic map (Analogy)?
> (3) Is it the case that a cognitive map is used *as if it* were a cartographic map (Metaphor)?
> (4) Is it the case that the cognitive map has no real connections with what we understand to be a map, i.e. a cartographic map, and is neither an explicit statement, analogy or a metaphor but rather an unfortunate choice of phrase: 'a convenient fiction?'…In effect just a hypothetical construct. (p. 3)

One notices in these competing viewpoints a faint echo of some of the issues surrounding ethno-classifications: to what extent are they genuine representations in people's minds, as opposed to artefacts of the analyst's methodology? It is, of course, possible that the four viewpoints are not entirely mutually incompatible, and that elements of each can be shown to be valid, depending on the investigatory tools being employed. Thus, (1) seems to be a reasonable viewpoint if we consider that so-called 'place cells' in the region of the brain called the hippocampus are preferentially activated in response to specific locations, and that the level of activation also signals the presence of familiar perceptual cues at that location (thus indicating episodic memory) ([195], see also [196] for a review). Viewpoints (2) and (3) appear to gain some validation from empirical studies such as that of Pocock [197], who showed that residents of a city tended to draw more interconnected, 'sequential' maps dominated by roads, whereas visitors and tourists could only produce 'spatial' maps, dominated by buildings, landmarks or districts scattered over the map, with incidental connections between map elements.[1] Finally, and partly in keeping with viewpoint (4), Gärling et al. [194] argue that:

[1] Pocock acknowledges the limitations of his findings, pointing out that "*the overall correspondence between possessed information and the elicited map is a function of graphicacy*", and that it may be unrealistic to assume equivalence between groups.

> The cognitive map is basically a semantic long-term memory representation of the properties of and the components in the environment that has been acquired through repeated encounters with that environment. The particular content of the cognitive map is adapted to its function and, if this function is to facilitate movement and travel in large-scale and medium-scale environments, it includes information about places (destinations for travel), spatial information, and travel instructions. (p. 25)

Linde and Labov [198] pioneered the use of small-scale mapping tasks in order to obtain a rich corpus of domain-specific spontaneous speech, with which to probe their consultants' spatial reasoning. They asked New York City residents to describe their apartments, as this subject, they argued, had the advantages of being general, well-motivated, controlled, objective, well-practiced, and occurring in a conventional social setting. They found strong regularities in the way their consultants described the layout of the rooms in their apartments: two major patterns dominated, namely the 'existential construction' (*...through an archway is the living room*) and the 'tour' (*...if you kept walking straight ahead, directly in front of you, you would find a bathroom*) (pp. 929–930). There were also strong correlations between the branching patterns of rooms (off a hallway) and the speaker's inclination to 'enter' it during the imaginary 'tour'.

The aims of the current chapter differ from the Linde and Labov study in four important ways. First, the scale of the spatial network investigated here is the entire B. R. Hills ecosystem, consisting of a mosaic of several habitat types, and covering over 300 km^2. The distances involved, and arguably also the complexity of the components in the spatial network, are naturally far greater than in the Linde and Labov study, and this may have a profound influence on the way people conceptualise such a network. Second, the New York study tried to uncover regularities in the way people described their apartments through the medium of language; the actual contents of the apartments were, in themselves, of little consequence. Here, I am interested in finding out not only how people talk about landscape and forest types, but also the encyclopaedic knowledge they possess about the characteristics of these places. Third, while the apartments of New Yorkers—at least the features of the apartments that would normally be described to a naïve listener—can be expected to remain more-or-less static, the forest that the Solega live in is a highly dynamic spatial network. Ecosystems are constantly changing entities, and the drivers of this change include the life cycles of plants and animals (i.e. the phenology[2] of these species), the migration of animals, the swelling and subsiding of water bodies, and the cycling of the seasons.

At least seven different tree phenological stages are recognized in Solega, on the basis of the appearance (or absence) of vegetative and reproductive structures on the tree; these include *baḷḷu* 'bare', *eḷa:ku* 'new leaves' (indicating a previously bare deciduous tree that has just begun to sprout new leaves), *ele* 'leaves' (indicating a foliage-covered tree), *muggu* 'buds', *hu:* 'flowers', *ka:yi*

[2] Phenology is the study of recurring plant and animal life cycle stages, or phenophases, such as leafing and flowering of plants, maturation of agricultural crops, emergence of insects, and migration of birds (USA National Phenology Network website).

'unripe fruit' and *haṇṇu* 'ripe fruit'. In addition, the intermediate stage *miḍi* 'fruit set' is also named in Solega, but as these very small fruit are quite hard to observe, this term is not commonly used in the context of tall rainforest trees. A few examples of this type of knowledge are shown in Table 5.1—the two consultants NG and MRM were asked to freely recall the names of as many forest trees as they could remember, and to describe their phenological status at the

Table 5.1 Excerpt from the author's field notes, showing NG and MRM's knowledge of forest tree phenology for the month of March

	Condition (26/03/09)						
Tree	*baḷḷu* (bare)	*eḷa:ku* (new leaves)	*ele* (leaves)	*muggu* (buds)	*hu:* (flowers)	*ka:yi* (unripe fruit)	*haṇṇu* (ripe fruit)
matti	✓						
bejja	✓						
honne		✓					
aranelli		✓					
ka:ḍageru	✓						
buḷuga		✓					
bu:ta:ḷe					✓		
i:raṇṇa	✓						
oluge	✓						
kagga:li	✓						
toṭṭi	✓						
anase			✓		✓		
soravilu		✓					
kara:va:di			✓		✓		✓
kallicci			✓				
biḷicci			✓				
beṭṭasuṇḍe			✓				
bu:ra			✓				
udi					✓		
kincaga					✓	✓	
araviḷu			✓		✓		
ke:silu			✓			✓	
torrema:vũ			✓	✓			
kaĩyebe:vu			✓		✓	✓	
mutta:ga			✓		✓		
kaḍaba	✓						
daḍasa:lu			✓	✓			
hatti							✓
ku:ḷi					✓		
kallumuttaga	✓						
gojja:li		✓					
sappe ambu			✓	✓			

time of questioning (the interview was carried out on 26 March, 2009). This was the start of the pre-monsoons, and many of the lowland trees were either still bare, or had just started to sprout new leaves. Although my consultants needed to confer with each other to arrive at a satisfactory consensus for some of the rarer trees, for the large majority of trees, they were able to provide near-instantaneous responses in agreement. During the walk back to my field station following this interview, I was able to verify that all the trees we passed along the way were indeed 'doing' exactly what my consultants had reported.

As the following description of landscape/forest terms makes clear, Solega people also associate each habitat with a particular suite of plant species. As each species is characterised by its own phenological cycle, it is possible that each habitat (i.e. forest/landscape) has a unique 'phenological signature' associated with it at each point in the seasonal cycle. Might these cyclic changes also form part of the Solega's mental representation of the landscape, and if so, how would the idea of a cognitive map need to be altered to accommodate temporal processes?

This chapter addresses many of the issues mentioned above, with respect to a suite of landscape terms in Solega. Not surprisingly, Solega has a rich vocabulary to describe the numerous landforms and water bodies that are commonly found on their ancestral lands—many of these are presented in Table 5.2 (see also Fig. 5.1). In the following section, I present transcripts of the interviews I carried out with two groups of Solega speakers, where I asked them specific

Table 5.2 Solega landscape and hydrological terms

Geological and biological landscape features	
adda:lu	Sheer drop
arre	Rock platform
eṇe	Flat land at the top of a hill or mountain
ogari	Cave
oḍagu	Sides of mountain; steeper than a *bo:re*
oḍḍu	Pile of boulders; rocky slopes
oḷḷe	Narrow gaps between boulders
o:ḍe	Crevice
õgu	Valley
kaṭṭe	Shore of a lake; artificial raised boundary around a lake
kaṇḍi	Boundary around a lake
kadiru	A grove of similar trees
ka:ḍu	Forest
koḷḷa	Steep slope
ko:be	Valley; flat land between two mountains
ko:sa:ru	Rocky terrain that is hard to navigate; found high up on mountains
gadde	Flat, marshy land
guḍḍe	Small hill

(continued)

Table 5.2 (continued)

Geological and biological landscape features	
goḍalu	The curve of the base of a hill or mountain
gonga:ru	Gorge; river valley with steep walls
go:re	Steep riverbank
baṇḍe	Boulder
bara bu:mi	Wasteland with only small plants and gravely substrate
bo:re (kaṭṭu)	Gentle slope
bo:ḷi	Summit of hill, usually covered with grassland
daddu	Deep vertical crevice in mountain rock
dimba	Flat land at an elevation, usually covered in forest
paḍuga	Flat land covered in bushes
su:ji beṭṭa	High pointy mountain peak
hoḍagu	Region halfway up a hill slope
kibbari	Region halfway up a hill slope
hottu	Riverbank
ho:ḍe	Rocky terrain with caves and ravines

Hydrological features	
aṇṇi	Pond or marshy area
abbi	Waterfall
aḷḷa	Stream
baraḷḷa	Dry riverbed
si:raḷḷa	Flowing stream
guṇḍi	Water hole, usually dug by hand
ba:vi	Well
maḍu	Large pool at the base of a waterfall
kaḍavu	Stream
kere	Artificial reservoir
kolli	Natural spring
koḷa	Water collecting at a natural spring, or at the base of a waterfall
ju:ḷi	Natural spring
dumuki	Waterfall
soṇe	Cavity on a rock platform that contains water

Miscellaneous	
(benda) kari	Ground that has been blackened by fire
hullu thoṭṭi	Land with clumps of grass that have not been burnt
ma:ḷa	Forest clearing, may be dominated by bushes
bailu	Forest clearing

Adapted from [166]

Fig. 5.1 Some common geological and hydrological features named in Solega. (**a**) *arre*, a rock platform, often used for washing or drying clothes, drying grain, sharpening machetes, etc. (**b**) *ko:sa:ru*, rocky terrain that is hard to navigate, (**c**) *aṇṇi*, boggy ground that serves as a perennial source of water, (**d**) *ba:vi*, a well, (**e**) *guṇḍi*, a temporary water hole, (**f**) *soṇe*, a cavity in an *arre* that sometimes holds rainwater

questions about various forest and landscape types. I mentioned the importance of contextualising empirical language data in Sect. 1.3, and in this respect, it is important to remember that the information presented in the following transcripts probably represents features of these forest and landscape types that would normally only be mentioned explicitly when addressing a naïve out-group member. To continue Linde and Labov's theme, a New Yorker might describe the layout of rooms in his new apartment to a friend only if that friend had never before been a guest there. A frequent visitor to the apartment who posed the question "*How's the apartment?*" might get a very different answer—s/he might be told that the apartment was a mess because of the owner's young children, that the owner's new dog had ruined a lot of the furniture, or that the kitchen was currently being renovated. The fourth way in which this chapter differs from the New York study is therefore the fact that the primary data tend more towards the 'elicited' end of the spontaneity spectrum, than towards the 'naturalistic'. I do not see this as a problem because, as mentioned earlier, I am equally interested in the encyclopaedic knowledge of the Solega, as I am in their means of communicating about landscapes.

5.2 *Ka:ḍu* as a Landscape

In addition to these terms that largely denote the non-living (geological and hydrological) environment, Solega also has a range of compounds that incorporate the word *ka:ḍu*, which is glossed in Table 5.2 as 'forest'. It was hinted in Chap. 3 that the gloss 'forest' is far too simplistic for this versatile word, as the compounds formed with *ka:ḍu* include labels not just for forests, but also for boulder fields, grasslands and fallow land. However, as the descriptions of these forest types (given below) illustrate, the topographic features of the landscape largely determine where they tend to occur. Many of the geological/topographical landscape terms shown in Table 5.2 form lexical compounds with *ka:ḍu* to generate an additional set of landscape terms that have highly variable characteristics. This topic was first investigated during the first few weeks of my fieldwork in the B. R. Hills. A group of young Solega men (their names have been abbreviated to JS, NG, JV in the following transcripts) were asked to first provide a list of terms used to describe the forest and different parts of the landscape. When a list of words was obtained, the men were asked to give any information they could, on the plant and animal life in these landscape types, along with any salient information that might help an outsider understand the meaning of these words. These Solega were reluctant to speak to me in their native language at this stage, preferring instead to use Kannada as the means of communication (their stated reason for preferring Kannada was that their Solega was not 'good enough').

In a later session, the same information was asked of an elderly Solega man (BG), who was by then quite familiar with my work, and was willing to speak on any topic in fluent Solega. At this interview, a younger Solega man (MRM) acted as interpreter and interviewer—I would ask him questions in Kannada, and he would, in turn, relay those queries to the main consultant (the elderly man) in conversational Solega. During such interviews, it was not uncommon for the younger man to be addressed directly by the consultant more often than I was (as determined by the direction of the latter's gaze, for instance). While I thus had less control over the direction of the interview, the advantage of this technique was that BG's speech was far more naturalistic; I could also be certain that the information being provided really did lie within the domain of Solega traditional knowledge. The interviewer was, after all, a Solega, who would have alerted me to the presence of any inconsistencies in the older man's speech.

The descriptions below are the result of combining the information from both elicitation sessions. In general, the data from both sessions were largely consistent, with the exception of a handful of landscape terms—these are further discussed below. Following the basic descriptions of all landscape types, I discuss how the Solega might perceive these very different landscapes as a patchwork of vegetation types, climatic zones and resource-producing areas. Most importantly, it will become apparent that when other aspects of Solega TEK are considered in conjunction with their conception of landscape/forest types, the latter comes to represent a dynamic cognitive map—one which is updated with the change of seasons to take into account the movements of wild animals and the availability of plant resources. Transcripts in which Kannada was the dominant language of communication have been labelled (K), while those in Solega have been labelled (S). Given the mixed composition of the groups, however (I as an English and Kannada 'speaker', and the consultants used to speaking a mixture of Kannada and Solega), many of the utterances are not in 'pure' code.

5.3 Landscape/Forest Types in Solega

All forest types in Solega can be loosely classified as either *male ka:ḍu* 'hill forest' and *na:ḍu ka:ḍu* 'plains forest'; *male* and *na:ḍu* incidentally, are terms are used by modern Kannada speakers as well. It is possible that these are recent Kannada borrowings into Solega (*tho:pu ka:ḍu* being the Solega equivalent of *male ka:ḍu*, and *(h)orega:ḍu* being the Solega equivalent of *na:ḍu ka:ḍu*, see below); however, given the large overlap between the Solega and Kannada lexica, this would be quite hard to prove. Nevertheless, these two words are quite interesting for the following reasons:

1) Even though Solega consultants would occasionally use the word *male ka:ḍu* when talking about 'their' forest, some equated it with the Solega term *tho:pu ka:ḍu* (discussed below), suggesting that there was no difference between the two.
2) *Male ka:ḍu* does indeed seem to be a superordinate term for a range of forest types that have names in Solega (see Extract I). The Kannada word '*male*' could be glossed as 'hill forest', and lacks the fine resolution of the other Solega landscape/forest terms described below.

Extract I (K)

(AS) *Male ka:ḍu andare?*
(What does '*male ka:ḍu*' mean?)

(JS) *Phullu male ka:ḍu*
(It's all *male ka:ḍu* (here).)

(JV) *Ellanu miks irutte*
(It's everything mixed up.)

(JS) *Ella miks irutte, phullu. Alli aḷḷa ka:ḍu baratte, koḷḷa baratte, go:re baratte, guḍḍaga:ḍu baratte, tho:pu ka:ḍu, arre ka:ḍu, i : male ka:ḍu madhyadalli ella ka:ḍu baratte.*

(It's everything mixed up, all of it. It includes river forests, valley forests, ridges, hill forests, *tho:pu* forests,[3] *arre* forests, these are all inside the *male ka:ḍu*.)
na:ḍu ka:ḍu ondu buṭṭu. Matte kutarega:ḍu buṭṭu, ellanue se:rutte.
(Except for *na:ḍu ka:ḍu*. And *kutarega:ḍu*. Apart from these, it's a combination of all the rest.)

(AS) *Pra:ṇigaḷu?*
(Animals?)

(JS) *Ã, male ka:ḍalli ja:sti pra:ṇigaḷu irutte.*
(Yes, there are a lot of animals in the *male ka:ḍu*)

(AS) *Matte caḷi?*
(And is it cold?)

(JS) *Illa, illa, caḷi matte bisilu eraḍue: irutte... ella pra:ṇi irutte alli.*
(No, no, it can be both cold and hot... all the animals live there.)

3) The word '*na:ḍu*', which can be glossed as 'district' or 'province' (as opposed to 'urban centre') in Kannada, is often used by the Solega to convey a sense of otherness. Thus, non-Solega (and hence Kannada-speaking plains people) are called *na:ḍavaru*, *-avaru* being a suffix derived from the third person plural pronoun, and meaning 'people associated with a particular feature, place or occupation'. In the following variant of the many *goruka* 'spider' songs that the Solega sing at festivals, the separation of Solega and non-Solega identity is stated quite unambiguously, through the use of *na:ḍu*.

ka:rañyana makka ka:ḍella uwe, ka:rañyana makkaelo ka:ḍallaelo,
ni:nu billañyana makka na:ḍellaelo, billañyana makka:ga na:ḍellaelo
ni:nu na:ḍu de:shella ninna:daiyo, na:ḍu de:shellawe ninna:delo
ka:ḍu go:ḍu ella namma:delo, ka:ḍu go:ḍu ellaelo namma:delo
The children of *Ka:rañya* are all (of) the forest (REPEAT),
You children of *Billañya* are all (of) the *na:ḍu* (REP.),
The whole of the *na:ḍu* country is yours (REP.),
The whole of the forest is ours (REP.).

4) Finally, when being asked to identify plants and animals from pictures in field guides, consultants would often indicate that they had never seen a particular organism by replying that it was from the *na:ḍu ka:ḍu*.

Gottilla, na:ḍu ka:ḍina adu
(I don't know, it's a *na:ḍu ka:ḍu* (bird).)

The above examples show that *male* and *na:ḍu* are labels for two contrasting macrohabitats. However, these labels are used not only in the biological domain, but also to mark an important social and cultural divide between Solega and non-Solega. In the following extract, recorded during a bird name elicitation session, BG and MRM refer to the *male ka:ḍu* using the possessive pronoun *namma* 'our', demonstrating a strong sense of affinity with the hill forests.

[3] To be described below.

Extract II (S)

(BG) *A:kaḍe hakki*
[Looking at the picture of an unfamiliar bird] (It's a bird from over there (*na:ḍu ka:ḍu*).)

(MRM) *Namma e:nadaluve kaṇḍu bandare he:ḷu, namma male ka:ḍina, tho:pu ka:ḍina ira:da.*
(Tell us if you happen to see anything of ours, from our *male ka:ḍu*, from our *tho:pu ka:ḍu*.)

(BG) *Ã adu ondu saṇṇadu gubisiyakki. Gubisiyakki namma ka:ḍina, male ka:ḍiga bara:du innue*
(Yes, that's a small sparrow. The sparrow is of our forest, it still comes to the *male* forest.)

Having looked at the major perceptual division in landscape/forest types made by the Solega lexicon, I will now proceed to describe, in some detail, the other terminology used by the Solega to describe their forest (Fig. 5.3).

5.3.1 Tho:pu ka:ḍu

As mentioned above, this first term is slightly problematic, as I was given two very different descriptions in the two elicitation sessions. In the second session, BG and MRM stated that *tho:pu* was merely another word for *male ka:ḍu*, and that in much the same way, the former also contrasted with *na:ḍu ka:ḍu*.

Extract III (S)

(BG) *Male kaḍe tho:pu. Na:ḍu ka:ḍu tho:pu ille. Adu na:ḍu ka:ḍe horrega:ḍu*[4] *endivẽ na:vu. Beṭṭa bo:ḷinda me:leka ira:du male ka:ḍu. Keḷageka ira:du na:ḍu ka:ḍu. Adavave na:vu horega:ḍu enna:du.*
(The *male ka:ḍu* itself is a *tho:pu*. The *na:ḍu ka:ḍu* is not a *tho:pu*. We call the *na:ḍu ka:ḍu* a *horega:ḍu*. Above the mountain grasslands, you get *male ka:ḍu*. Below, you get *na:ḍu ka:ḍu*. That's the one we call *horega:ḍu*.)

(BG) *Idu tho:pu ka:ḍu. Na:vu ira:della uwe i: ka:ḍella pu:ra tho:pu ka:ḍu.*
(This is *tho:pu ka:ḍu*. This whole place we're in, this forest is all *tho:pu ka:ḍu*.)

(AS) *I: ka:ḍu ma:tra athava…?*
(Just this forest (here), or…?)

(MRM) *Illa, ella se:ri. Idu namma BRT adella, idu ishṭu ma:tra tho:pu ka:ḍu. I:ga na:vu monne alliga ho:gidive˜lla. Po:ḍugaḷige, allinda keḷake ella horega:ḍu.*
(No, all of it together. This BRT of ours, right, it's only *tho:pu ka:ḍu*. The place we went to a few days ago, right, to the villages, looking down from there it's all *horega:ḍu*.)

[4] See *orrega:ḍu* below

Fig. 5.2 The major *ka:ḍu* types in Solega. (**a**) *tho:pu ka:ḍu*, a continuous expanse of highland forest; (**b**) *beṭṭa ka:ḍu*, 'mountain forest', consisting of a variety of vegetation types; (**c**) *doḍḍa ka:nu ka:ḍu*, cold, dark evergreen forest dominated by large trees; (**d**) *orrega:ḍu*, vegetation from the lower mountain slopes; (**e**) *kutare ka:ḍu*, plains vegetation, dominated by dry scrub; (**f**) *oṭṭuga:ḍu*, recently abandoned farmland; (**g**) *eḷa:vuga:ḍu*, land that has been fallow for a few seasons; (**h**) close-up of *beṭṭa ka:ḍu*, 'mountain forest'; (**i**) *aḷḷa ka:ḍu*, vegetation accompanying a

Fig. 5.2 (continued) stream in the lowlands, seen as a line of lusher green (*two left arrowheads*) terminating in a rockpool (*right arrowhead*); (**j**) *aḷḷa ka:ḍu*, vegetation growing by a larger mountain stream (currently dry); (**k**) *hoḍe ka:nu*, evergreen forest growing on the sides of a valley; (**l**) *saṇṇa ka:nu*, patches of high-altitude shola forest (*arrowhead*) surrounded by grassland; (**m**) *saṇṇa ka:nu*, view from within a shola forest, showing trees encrusted with mosses and other epiphytes; (**n**) *oḍḍuga:ḍu*, a boulder field; (**o**) *beṭṭa bo:ḷi* or *bo:ḷiga:ḍu*, a bare hill top covered with grassland; (**p**) *guḍḍega:ḍu*, vegetation growing on small hills.

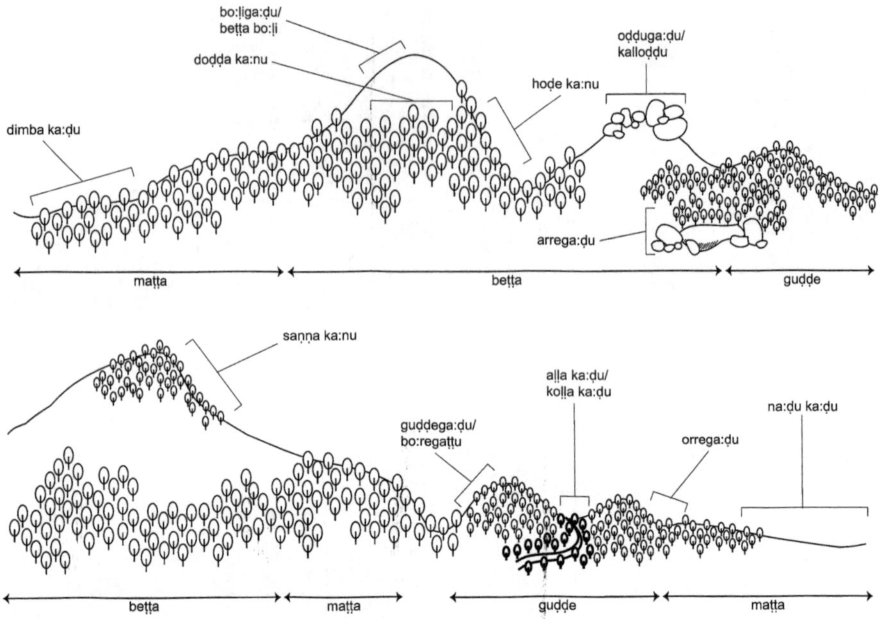

Fig. 5.3 Selected forest and landscape terms in Solega. Previously published in [8]

…

…

(BG) *Male ka:ḍinalli bisilu ille. Eshṭo: tampa:girtade.*
(It's not hot in the *male ka:ḍu*. It's quite cool.)

(BG) *Adakka tho:pu ka:ḍu. Tho:pu ka:ḍu—beṭṭa, guḍḍe, maṭṭa...*
(That's a *tho:pu ka:ḍu. Tho:pu ka:ḍu* includes mountain, hill and flatland forests…)

(MRM) *A:mele ka:nu ka:ḍu.*
(And then evergreen forests.)

In the above extract, both BG and MRM equate *male* and *tho:pu* on the one hand, and *na:ḍu ka:ḍu* and *horega:ḍu* (a variant of *orrega:ḍu*, below) on the other. MRM's comment halfway through the extract, as well as BG's final statements reinforce the idea that *tho:pu* is a kind of superordinate category, as is *male*—it seems to encompass everything that the Solega think of as 'their' forest.

The younger Solega men interviewed in the first elicitation had a radically different concept of *tho:pu ka:ḍu*. In the following extract, they all agree that it is a much smaller feature of the landscape:

Extract IV (K)

(JV) *Tho:pu andare ondu gru:p irutte.*
Tho:pu means a group.

(AS) *Maragaḷa gumpu?*
A group of trees?

(NG) *Ã, maragaḷa gumpu. A: gumpu maragaḷadu ashṭe.*
(Yes, a group of trees. That group consists of trees, that's all.)

(JS) *Adu ambugaḷu, matte ja:sti thik, tumba maragaḷu. Gumpu irutte, arda e:kare, ondu e:kare.*
(There are vines, and trees growing close together. It's a group, half an acre or an acre (in area).)

(AS) *Andare, tumba saṇṇadu...*
(So it's a very small...)

(JS) *Ã, solpa cikkadu. So:la fa:resṭu tara ka:ṇatte*
(Yes, it's quite small. It looks like a shola[5] forest.)

(NJ) *Gumpu mara. ondu, hattu mara idda:ga, adu tho:pu mara.*
(A group of trees, when you have about ten trees, that's a *tho:pu* of trees.)

(JV) *Illi, namma a:fis mum-bha:ga tho:pu ide*
(There's a *tho:pu* here, in front of our office.)

Here, the consensus seems to be that *tho:pu ka:ḍu* refers not to the entire hill forest macrohabitat, as BG and MRM assert, but rather to small, isolated groups of trees. It is quite possible that BG's conception of *tho:pu ka:ḍu* represents an older Solega meaning of the term—younger Solega, especially those who have had greater contacts with Kannada speakers, might generally be unaware of this meaning, and their understanding of the term may have converged with the standard Kannada meaning of *tho:pu,* which is 'orchard' or 'grove' (as in *ma:vina tho:pu* 'mango orchard' or *tengina tho:pu* 'coconut plantation').

5.3.2 *Beṭṭa:ga:ḍu*

In Extract III, BG mentioned that *beṭṭa:ga:ḍu* 'mountain forest' was one of the habitat types found within the *tho:pu ka:ḍu* macrohabitat. *Beṭṭa:ga:ḍu*, in turn, seems to be a grouping of many vegetation types, all of which happen to share the key feature of being found on a mountain.

Extract V (K)

(JS) *Doḍḍa beṭṭa:gaḷu, avella beṭṭa:ga:ḍu anta baratte. Alli doḍḍa ka:nu*[6] *baratte, saṇṇa ka:nu baratte, matte beṭṭa bo:ḷi, avella a:girutte, adakke beṭṭa:ga:ḍu. matte beṭṭa:ga:ḍinalli ḍra:i fa:resṭnu: ide. Ella maragaḷu ive, ja:sti.*

---

[5] See *saṇṇa ka:nu ka:ḍu* below

[6] These terms are explained below.

((The forest on) the big mountains are all called *beṭṭaga:ḍu*. You find evergreen forests there, shola forests, grassland, it's all there, that's why it's called *beṭṭaga:ḍu*. There's even dry forest there. There are lots of trees of all types.)

Based on the JS's stament, and on the discussions to follow, it would be inaccurate to suggest that *doḍḍa ka:nu*, *saṇṇa ka:nu*, *beṭṭa bo:ḷi* or even 'dry forest' are all 'kinds of' *beṭṭa:ga:ḍu*. Instead, *beṭṭa:ga:ḍu* should be thought of as a category that cross-cuts all the others; thus, while *doḍḍa ka:nu* 'evergreen forest' can occur both on mountains and on flat land, only that part of the evergreen forest that grows on a mountain slope is subsumed under the label *beṭṭa:ga:ḍu*. The same can be said of the 'dry forest' that JS mentions.

5.3.3 (Doḍḍa) ka:nu ka:ḍu

The term *ka:nu ka:ḍu*, which is more properly referred to as the *doḍḍa* 'big' *ka:nu ka:ḍu*, is usually glossed by biologists visiting the region as an 'evergreen forest'. The following is an extract from the second landscape elicitation session; BG is asked by MRM and the author to describe *ka:nu ka:ḍu* in as much detail as possible:

Extract VI (S)

(BG) *Doḍḍa giḍa ja:ga:dalli...adu ka:nu ka:ḍu. Beṭṭa:da me:le ka:nu beḷeda:du, caḷi ja:sti alli.*
(Where you find big trees…that's the *ka:nu ka:ḍu*. It grows on mountains, it's very cold there)

(MRM) *Kambaḷi gadde ka:nu endale doḍḍa ka:nu namaga hesaruva:si anta ka:nu. I: buraḍe benga:liye la:sṭ…adu la:sṭ endale, heccu kaḍime allinda eṇṭu kilomi:ṭar hattu kilom:ṭar tanaka biri (ka:nu ka:ḍu irutte).*
(*Kambaḷi gadde ka:nu* is a *doḍḍa ka:nu*. That's our prototypical *ka:nu*. This <PLACE NAME?> is the last (furthest extent). From there you only get *ka:nu ka:ḍu* for about eight or ten kilometres.)

(AS) *Ella onde: ka:ḍu?*
(Is it all a single forest?)

(MRM) *Onde: ka:ḍu..alli ḍiva:iḍ irutte. Idu kambaḷi gadde ka:nu, idu bandu illi… gombegallu ka:nu, doḍḍa sampige ka:nu, cikka sampige ka:nu, i: tara alliye ḍiva:iḍ a:girutte, ka:ḍu ka:ḍu hesaru. Andare so:la beḷedira:du, alliye doḍḍadu.*
(It's a single forest…but divided up. You have *kambaḷi gadde ka:nu*, and then…*gombegallu ka:nu, doḍḍa sampige ka:nu, cikka sampige ka:nu*, it's divided up like that, into different forest names. So where the *so:la* grows, that's *doḍḍa ka:nu*.)

(AS) *Matte alli ma:muli pra:ṇigaḷu sigutte?*
(And you find the usual animals there?)

(BG) *Ka:nina oḷage kaḍave ade, ka:ḍemme ade, matte bekku ade, kotti ade... ko:tigaḷu irtave... a:ga matte ku:randi, muḷḷandi ade, matte handi ade... ishṭella ka:nina oḷage ade. Adu ka:ninalliye ira:ku. I: ka:ḍiga bandale barakka sa:dya illa.*
(Inside the *ka:nu* you find sambhar, gaur, civets, there are monkeys, and then mouse deer, porcupines, pangolins, all these are inside the *ka:nu*. The have to remain inside the *ka:nu*. It's not possible for them to come to this [open, dry] forest.)

This extract, combined with information from the first elicitation session, shows that the label 'evergreen forest' fails to capture the extensive ecological knowledge that the Solega have about the entity called *ka:nu ka:ḍu*. 'Evergreen forest' simply does not take account, for example, of the fact that "few animals (mammals) like to live here", and that the ones that do, "leave during the rainy season, because of the cold". The Solega also avoid living in these parts, for much the same reasons. Solega perceptions of the *ka:nu ka:ḍu* ranged from the mundane and unsurprising, ("it is always cold and dark here, because the sunlight doesn't reach all the way to the ground", or "there are rivers here with water year-round, and so the forest is always green"), to detailed species lists of the flora that can be found in this forest type:

Extract VII (S)

(BG) *Ka:ninalli ha:le mara ade, thuruve mara ade, kakkilu, bikkilu, bella:ḍe, ku:ma:ũ, kende, soravilu, aravilu, hebbe:u... aravilu kende maradalli je:na:de.*
(In the *ka:nu*, there's the *ha:le* (?) tree, the *thuruve* tree (*Ligustrum perrottettii*), etc…there are bees/honey on the *aravilu* (?) and *kende* (*Elaeocarpus tuberculatus*) trees.)

(MRM) *Nalavattu aivattu je:nu baratte onde maradalli.*
(Forty or fifty colonies come to a single tree.)

(BG) *Ne:ri, ne:ri mara,a:ga matte, i: ta:re mara ondu ille. A ka:nu oḷage e:va ja:ga:dalli uve ta:re mara huṭṭa:dille. A:mele, koḷamada mara, mi:nada mara...*
(The *ne:ri* tree, and then, the *ta:re* (*Terminalia bellirica*) tree is the only one missing. There isn't a single *ta:re* tree growing anywhere inside the *ka:nu ka:ḍu*. And then, the *koḷama* (?) tree, the *mi:na* tree (*Myristica dactyloides*)…)

As this excerpt shows, BG not only names several species of tree that are typical of a *ka:nu ka:ḍu*, but also informs me about a particular species (the *ta:re*) that is never found in this forest type. This information was offered spontaneously, and shows that for BG, at least, the concept of *ka:nu ka:ḍu* is a very precise one.

5.3.4 (H)orrega:ḍu (also na:ḍu ka:ḍu, kutare ga:ḍu)

Immediately following the list of *ka:nu ka:ḍu* trees, BG went on to contrast that group of species with those found in the second forest/landscape type, namely *orrega:ḍu*. This suggests that the distinction between *ka:nu ka:ḍu* and *orrega:ḍu* might be a major perceptual one, and that (although it was not explicitly stated by my consultants) it may be possible to group other forest types under these superordinate categories. In the following extract, BG directly compares the two forest types, providing one key criterion for distinguishing between them:

Extract VIII (S)

(BG) *Orrega:ḍu, begga:ḍu, mara kaḍame irta:della, a: ja:ga orrega:ḍu. mara kaḍame adella, i: ja:ga solpa mara illa, ja:sti illa mara... ella ondu ondu irtade.*
(*Orrega:ḍu*, or *begga:ḍu*, that's where you find fewer trees. You find fewer trees, right? Trees are a bit scarce in this place, there aren't many trees… they grow scattered here and there.)

(AS) *Cikka giḍa ja:sti irutta?*
(Are there many small plants?)

(BG) *Cikka giḍa ja:sti...doḍda giḍa ja:ga:dalli, adu ka:nu ka:ḍu.*
(There are many small plants… where you find big trees, that's the *ka:nu ka:ḍu*.)

Despite the scarcity of large trees, BG is nevertheless able to name several species that might be found in an *orrega:ḍu*. It is noteworthy that the overlap in tree species between the following list and that provided for the *ka:nu ka:ḍu* in Extract VI is minimal. In fact, out of a total of 24 distinct trees named in the two extracts, only *soravilu* (*Acrocarpus fraxinifolius*) and *hebbe:vu* (*Toona ciliata*) were listed as occurring in both forest types:

Extract IX

(BG) *Orrega:ḍinalli e:na:de? Bejja:da mara ade. Gejja, matti, bi:ḍi, soravilu, hebbe:u, be:nada mara, matte aĩle mara, aḍali mara, a:ga matte nelli, aṇṭuva:ḷada mara... aṇṭuva:ḷada marada ka:yi ma:ḍtivĩ allava? A:ga matte innu: sele mara, ba:ge mara...*
(What's there in the *orrega:ḍu*? There's the *bejja* tree (*Anogeissus latifolia*), *gejja* (?), *matti* (*Terminalia crenulata*), *aṇṭuva:ḷa* (*Sapindus laurifolius*)… we eat the fruit of the *aṇṭuva:ḷa* tree, don't we? And then also the *sele* tree (*Albizia odratissima*), the *ba:ge* tree (*Albizia amara*)…)

…

…

(BG) *A: ja:ga:dalli suṇḍrega:ḍi, si:ge ambu, magani be:ru, matte sucilu giḍa, uruvilu mara, ishṭella be:ka:dashṭu horrega:ḍina:du. Ashṭella illi huṭṭa:dille.*

(In that place, you get *suṇḍrega:ḍi* (?), *magani* vine (?), and the *sucilu* plant (?), the *uruvilu* tree (?), all these belong to the *horrega:ḍu*. They don't grow here.)

(MRM) *Illi uruvilu mara illevelella?*
(There's no *uruvilu* tree here, right?)

(BG) *Illi ille. Ge:ru mara pu:ra i:kaḍe...*[7]
(Not here. There are lots of *ge:ru* trees (*Semicarpus anarcardium*) this way.)

When the consultants in the first elicitation session were asked to describe *orrega:ḍu*, they described traits that were different, but not inconsistent with those mentioned by BG. In the following extract, JS and NG emphasise the topographic qualities of *orrega:ḍu*.

Extract X (K)

(JS) *Alli ful maṭṭa baratte ondu ka:ḍu, ful maṭṭa, alli beṭṭa sṭre:ṭ a:girutte. A: maṭṭa ka:lu ba:ga, arda ba:ga, i: beṭṭa arda ba:ga tagoṇḍu orrega:ḍu baratte... i:ga maṭṭa bandu beṭṭa irutte hi:ge, allava? Adaralli madya bandu ma:tra orrega:ḍu baratte.*
(There you see a flatland forest, completely flat, then a hill goes straight (up. A forest that's) a quarter or half on flat land, and half on a hill is an *orrega:ḍu*... It's flat here, and then the hill goes like this, right? Only in the middle of that do you get *orrega:ḍu*.)

(NG) *Idu maṭṭaga:ḍiga ue borega:ḍiga ue eraḍaka ue se:ri, adakka orrega:ḍu.*
(When the flatland forest and the hill-slope forest join, that's *orrega:ḍu*.)

It is quite likely that the terms *na:ḍu ka:ḍu*, *kutarega:ḍu* and *orrega:ḍu* are used interchangeably in some contexts, as they were often substituted, or explicitly equated, in the two elicitation sessions. In Extract III, for example, BG states that *na:ḍu ka:ḍu* and *orrega:ḍu* are much the same, also adding that this was the name for the forest below the hill grasslands—this matches the description in Extract X, where *orrega:ḍu* is said to be the forest at the foot of a hill and on the adjacent flat land. In the following extract, when asked to explain *kutarega:ḍu*, BG states clearly that this term denotes the same forest type as *orrega:ḍu* (which, as discussed above, was said to be identical to *na:ḍu ka:ḍu*). Consistent with this, MRM describes the vegetation in such areas as consisting of small trees, while BG suggests that the grass that grows there is typical *na:ḍu ka:ḍu* grass.

Extract XI (S)

(BG) *Orrega:ḍu kutarega:ḍu. Ittaka na:vu oitivĩ keḷeka, adu kutarega:ḍu. Ittakella, maṭṭaga:ḍu, guḍḍega:ḍu, ishṭakella biri tho:pu ka:ḍe.*
(An *orrega:ḍu* is a *kutarega:ḍu*. Like we said, downhill from here it's *kutarega:ḍu*. All this, the flatland forest, the hill forest, all this is just *tho:pu ka:ḍu*.)

[7] BG goes on to warn about the fruit of the *ge:ru* tree: the flesh is edible, but the seed is poisonous.

(MRM) *Kutarega:ḍu endare na:ḍu ka:ḍu e:ria.*
(*Kutarega:ḍu* means the *na:ḍu ka:ḍu* area.)
(BG) *Rangappana beṭṭa:da keḷaka:della, adu kutarega:ḍu*
(At the bottom of Rangappa mountain, that's *kutarega:ḍu.*)
(MRM) *E:kandare alli cikka cikka maragaḷu irtave.*
(Because small trees grow there.)
(AS) *Hullu ja:sti irutte, alli?*
(Is there a lot of grass there?)
(BG) *Na:ḍu ka:ḍina hullu ade—uṇuga:na hulle, a:mele jabbara hullu.*
(Just *na:ḍu ka:ḍu* grass—*uṇuga* grass, then *jabbara* grass.)
(AS) *I:ga, na:ḍu ka:ḍu matte kutarega:ḍu onde?*
(So, are *na:ḍu ka:ḍu* and *kutarega:ḍu* the same thing?)
(BG) *Na:ḍu ka:ḍu kutarega:ḍe. Na:ḍu ka:ḍu endale beṭṭa:dinda keḷeka. Beṭṭa:dinda keḷeka kutarega:ḍu. Beṭṭa:dinda me:leka maṭṭaga:ḍinalli, adu tho:pu ka:ḍu.*
(*Na:ḍu ka:ḍu* is *kutarega:ḍu. Na:ḍu ka:ḍu* means (forest) below a mountain. Below a mountain, you get *kutarega:ḍu*. The flatland forests on the mountains, that's (a part of) *tho:pu ka:ḍu.*)
(AS) *I: kutarega:ḍu matte orrega:ḍu, ye:n antara ide?*
(This *kutarega:ḍu* and *orega:ḍu*, what's the difference?)
(BG) *Ade, kutarega:ḍu enna:du, orrega:ḍu enna:du.*
(They're the same, it can be called *kutarega:ḍu* or *orrega:ḍu.*)

Critically, BG makes a clear distinction in Extract XI between flatland forests (*maṭṭaga:ḍu*) that occur at elevation, and those that can be found near the transition from mountains to plains. The clear differences in altitude, along with accompanying changes in vegetation serve to distinguish the former (*dimbaga:ḍu*, see Extracts XXVI and XXVII below), which is considered part of the *male/tho:pu ka:ḍu* macrohabitat, from the latter (*orrega:ḍu/kutarega:ḍu*), which are thought to be equivalent to *na:ḍu ka:ḍu*.

The descriptions of the landscape/forest terms discussed so far show that these forest/landscape terms can either be primarily described on the basis of species composition, or on the basis of topographic properties. In addition, speakers talk about some forest/landscape words in terms of human interactions, as exemplified by *oṭṭuga:ḍu*, *eḷa:vuga:ḍu* and *begga:ḍu*.

5.3.5 *Oṭṭuga:ḍu/eḷa:vuga:ḍu*

Simply put, *oṭṭu ga:ḍu and eḷa:vu ga:ḍu* describe stages in the regeneration of farmland, when it is abandoned after 2–3 years of cultivation. Such 'fallow land' may still have some utility, as illustrated by the following:

Extract XII (S)

(BG) *I:ga na:vu hola ma:ḍidivãlla, i:ga na:vu mutkadgaddeli hola ma:ḍive˜. I:ga na:vu hindeli ma:ḍittu, illi, kaḷḷare beṭṭadalli. Kaḷḷare beṭṭadalli ma:ḍida:ga, adu buṭṭu banda me:le e:n a:gottu, adu (h)oṭṭuga:ḍu. Buṭṭa:kina bandu uṭṭõlla, a:seka adu eḷa:vuga:ḍu. buṭṭanta oṭṭuga:ḍu adu. A:ga oṭṭuga:ḍinalli e:n a:gottu, adu barta, barta, barta, e:n a:gottu, adu barta, barta, barta, mara, maṇḍi giḍa ella beḷedaru, beḷedaru bartira:du.*

(Now we've farmed some land, right? A farm in Mutkadgadde, for example. We used to do that in the past as well, on Kallare Betta. After farming some land there, when you leave it and go away, what happens—you get *oṭṭuga:ḍu*. You leave the land and go away, right? First you get *eḷa:vuga:ḍu*, and then *oṭṭuga:ḍu*. And what happens in an *oṭṭuga:ḍu*? All sorts of plants and trees keep sprouting and growing in it.)

(MRM) *Adakka oṭṭuga:ḍu.*

(That's an *oṭṭuga:ḍu*.)

(BG) *Adu oṭṭuga:ḍu. Eḷa:vu—"No:ḍru ba, neḍi ba:, geṇasina ambu i:teno neḍi no:ḍru ba:rõ"... "Kalluwe pu:ra, bari maṇṇuwe, pa:ille, geṇasu a:ga:dille," A: taradalli na:vu alli ma:ḍida holaka ho:ku na:vu. A: (h) oṭṭiga. A:ga: ho:da:ga alli geṇasu. A: ja:ga:dalli je:nu sikka:dille. Hola ma:ḍa idda ja:ga:dalli je:nu sikka:du. E:kendale hola ella ma:ḍa:gidella, ella a:ḷa:della.*

(That's an *oṭṭuga:ḍu*. As for *eḷa:vu*[8]—"Go have a look, come back and tell me if there are yams there or not." … "It's all rocks and dirt, no good, you can't get yams there." That's when we'd go the fallow land. You find yams when you go to the *oṭṭuga:ḍu*. You don't find honey in those places. You only get honey from land that hasn't been farmed. It's because you've made a farm, and there are people everywhere, right?)

It was only when MRM explained some crucial points regarding these two forest/landscape terms (during a later transcription session) that the significance of this rather confusing extract became clear. According to MRM, farmland that has very recently been abandoned is called *eḷa:vuga:ḍu*; after a period of several months, when this plot of land starts to be recolonised by all manner of local plant species, it can be called an *oṭṭuga:ḍu*. About a year after the cessation of farming, the *oṭṭuga:ḍu* can be a good source of yams; honey, on the other hand, is rarely found in such places, as the Solega tend to choose treeless clearings for farmland. As these plots are often quite small, however (practical considerations usually force a household to clear and cultivate only as much land as can be tended by two adults), it would be unsurprising to find bee colonies on trees in the close vicinity of an *oṭṭug:a:ḍu*.

[8] This is a discursive strategy frequently used by BG: the representation of an event or a process by means of a mini-conversation between him and an unspecified partner, or group of partners.

5.3.6 Begga:ḍu

This forest/landscape type is characterised by its lack of utility to the Solega. Although BG did mention *begga:ḍu* in connection with *orrega:ḍu* (see Extract VIII above), he most probably intended to group these two forest types together (and in turn, contrast them with *ka:nu ka:ḍu*) only because they shared an important feature—the absence of big trees. This is not inconsistent with the description of *begga:ḍu* given in the elicitation session with younger speakers:

Extract XIII (K)

(NG) *Begga:ḍu, ella ḍra:i enta ful oṇa:gi iruvanta ja:ga istaḷa. Alli* e:nue be:re asaru giḍagaḷu illa.
(*Begga:ḍu*, it's a place where everything is dry. There are no other green plants there.)

(JS) *Begga:ḍu andare ye:n sigalla, je:nu sigalla, matte pa:se sigalla.*
(*Begga:ḍu* means you don't find anything there, no honey, no lichen.)

(AS) *Alli ya:va tara:da giḍagaḷu sigutte?*
(What sort of plants do you find there?)

(JS) *Mara du:ra du:ra irutte.*
(The trees are spread out.)

(JV) *A:mele alli ya:vadu: sigalla. Ni:r sigalla, pa:se, nellika:i, je:nu tuppa.*
(And you don't find anything there, no water, lichen, amla fruit, honey.)

(JS) *Adakke begga:ḍu andare alli ye:n sigalla, adu we:sṭ ka:ḍu ante:ḷi namma jana he:ḷta:re.*
(That's why *begga:ḍu* means there's nothing there, our people call it a 'waste' forest.)

(NG) *Namma avara ba:sheli be:ḍe ka:ḍu ante:ḷi.*
(In the language of our people it is called *be:ḍe ka:ḍu*.)

There are indications that the term *begga:ḍu* might actually be a 'functional' landscape category that cross-cuts several other basic forest/landscape categories. *Begga:ḍu* would then simply mean 'forest/land of any description which does not yield any useful products'. The following statement by JS supports this point of view:

Extract XIV

(JS) *Begga:ḍalli guḍḍega:ḍu ide, oḍḍuga:ḍu ide...*
(In a *begga:ḍu* you can find hill forests, boulder fields…)

The terms *guḍḍega:ḍu* and *oḍḍuga:ḍu*, which I have translated as 'hill forest' and 'boulder field' are lexicalised landscape terms in their own right (see explanations below). Still, JS states that they can be part of a *begga:ḍu*—this suggests that in this utterance, he is drawing attention to the fact that even other forest/landscape types, which do not yield honey, edible fruit or other tradable goods, can be labelled *begga:ḍu*.

5.3.7 *Hullu thoṭṭi, kari/benda ka:ḍu*

Like *oṭṭuga:ḍu* and *eḷa:vuga:ḍu*, this is another example of a landscape type created as a direct consequence of a human activity, namely burning in this case. Recognising the unpredictable nature of fire, the Solega have names for parts of a grassland that are or are not burnt by a fire.

Extract XV (S)

(BG) *Hullu thoṭṭi endale, i: ka:ḍu bei.tidda:ga, ka:ḍu bendu bendu, pa:ḍiga adara ishṭakave, bendu bendu ba:ṇe hullu, na:ralu hullu, ko:ḷi hullu, sã:ya, matte i: jabbar hullu, matte i: hullella gu:ḍu maḍa:gi ku:tu e:ṭinalli kiccu bidda:ga baga:rendu oitittalla. A:ga be:ya ja:ga:dalli bendoi.tittu. Be:ya idda ja:ga:dalli be:ya ille. Adava hullu thoṭṭi endivẽ. Benda ja:ga kari, benda ka:ḍu.*

(*Hullu thoṭṭi* means, when you light a fire in the forest, the fire spreads through the forest on its own. It burns all the *ba:ṇe* grass, *na:ralu* grass, *ko:ḷi* grass, ferns, and the *jabbar* grass, and because these grasses grow in clumps, the fire burns through them really fast. The areas that the fire went through are burnt, The places that didn't have fire are not burnt. These (last) are called *hullu thoṭṭi*. The burnt places are called *kari*, or *benda ka:ḍu*.)

5.3.8 *Aḷḷa ka:ḍu/koḷḷa ka:ḍu*

The many forest streams that flow through Solega territory are well known to the people living there, regardless of whether they are seasonal or perennial. The forests that grow alongside these watercourses are designated by the terms *aḷḷa ka:ḍu* and *koḷḷa ka:ḍu*—in Kannada, the words *aḷḷa* and *koḷḷa* can be glossed as 'stream/rivulet' and 'a deep place, a depth, the cleft in a rock, a cave, etc.' [199] respectively.

Extract XVI (K)

(AS) *Aḷḷa ka:ḍu andare?*
(What does *aḷḷa ka:ḍu* mean?)

(JV) *Sṭre:ṭ nadi ho:gtade.*
(A river goes straight (through it).)

(AS) *Nadiya pakkadalli idda ka:ḍu?*
(It's the forest next to a river?)

(JS) *Ã, solpa miks irtade fa:resṭ adu. Evargri:n tara miks. Aḷḷadalli mara:gaḷu ḍra:i a:galla. Ella tara maragaḷu ful gri:n irutte ja:sti. Adu gotta:gatte, akasma:t na:vu ondu bo:ḷi me:le ho:dare, ondu beṭṭa:da me:le ho:dare, adu*

pakka gotta:gatte, ella gri:n iruttade. Adu aḷḷa hi:ge beṇḍa:gi ho:gade. Solpa gri:n irutte. Adakke wa:ṭar sapla:i ja:sti idarinda.
(Yes, that forest is a bit mixed. Like the evergreen forest. The trees don't get dry next to a river. All types of trees stay green mostly. You can tell, when you happen to reach a bare hilltop, when you go up a hill, you can tell that (there's a patch of) green nearby. That's the river going around a bend like this. It's a bit green. It's because there's a greater water supply.)

Extract XVII (S)

(AS) *Koḷḷa ka:ḍu andare ye:nu?*
(What does *koḷḷa ka:ḍu* mean?)
(JY) *Se:ne a:ḷa ade. I:cari beṭṭa, a:cari beṭṭa, madyada ko:sa:ru, adakke koḷḷa*
(It's really deep. A mountain on this side, a mountain on that side, rocky in the middle, that's a *koḷḷa*.)

At first glance, the terms *aḷḷa ka:ḍu* and *koḷḷa ka:ḍu* do not seem particularly different in terms of their real-world referents: both seem to indicate what an English speaker might readily gloss as a 'riverine forest'. However, it is possible that the former term is preferred when describing the flatland forest growing on either side of a meandering stream. The presence of water in a restricted area, i.e. the actual watercourse, would cause a narrow strip of vegetation growing immediately on either side of the watercourse to appear more lush than the plant life further away. *Koḷḷa ka:ḍu*, on the other hand, would be used to describe the vegetation growing on the steep slopes of a deep gorge or river valley, i.e. the relatively narrow space formed by the sides of two adjacent mountains or hills, at the bottom of which might be a stream. This definition is supported by the respondents in Extract XVIII, who also note that the slope in a *koḷḷa* is far more steep that that of a *bo:re* (see section on *bo:re ka:ḍu*, which also refers to a forest growing on sloping land).

Extract XVIII (K)

(JS) *Koḷḷa andare ful ḍa:un adu... ḍi:p a:girtade. Ful i: tara... aḷḷa irutte, eraḍu kaḍe beṭṭa ful irutte.*
(*Koḷḷa* means it goes straight down…it becomes very deep, like this. There's a river, there are mountains on both sides.)

…

…

(AS) *Bo:re ka:ḍu matte koḷḷa ka:ḍu, ye:n antara ide?*
(What's the difference between *bo:re ka:ḍu* and *koḷḷa ka:ḍu*?)
(JS) *Koḷḷa be:re, bo:re andare solpa bo:re. Guḍḍega:ḍinalli baratte, no:ḍi.*
(*Koḷḷa* is different, *bo:re* means a slight slope. You can see it in hill forests)
(NG) *Koḷḷa andare, idu aḷḷa, koḷḷa, eraḍu i: tara se:rutte*
(*Koḷḷa* means, the river, the sides of the gorge, both meet like this.)

5.3.9 *Hoḍe ka:nu*

The above two terms therefore differentiate between riverine forests on the basis of the flatness of the land on which they are found. In the second elicitation session, however, BG volunteered the term *hoḍe ka:nu*, immediately after having defined the words *doḍḍa ka:nu* and *saṇṇa ka:nu* (see below for *saṇṇa ka:nu*). The presence of the morpheme *ka:nu* in *hoḍe ka:nu* indicates that this forest type is conceptually linked to the high-altitude evergreen and shola forests. Beyond this, it is difficult to determine from the available data (Extract XIX) what the differences, if any, are between *hoḍe ka:nu* and *koḷḷa ka:ḍu*. Unlike *koḷḷa*, the Kannada word *hoḍe* can be glossed as 'border, edge…' etc. [199], and it is possible that *hoḍe* forests are preferentially mentioned in discourse in which the actual 'body' of a mountain is the topic, whereas the term *koḷḷa* forest is used more when the low-lying region between two mountains needs to be highlighted. At this stage, however, my data suggest that these two terms are nearly synonymous.

Extract XIX (S)

(BG) *Idondu beṭṭada no:ḍdale sariya:da hoḍeve. Ittaka hoḍeve, attaka hoḍeve, ettaka ho:ga:ge uwe da:ri ille. Adakka hoḍe ka:nu.*
(If you look at a mountain, there's clearly a *hoḍe*. There's a *hoḍe* on one side, there's a *hoḍe* on the other side. You can't go forward on either side. That's a *hoḍeka:nu*.)

(AS) *Hoḍe andare?*
(What does *hoḍe* mean?)

(MRM) *Hoḍe ka:nu endare, i:ga madyadalli doḍda sampige aḷḷa baratte.*
(*Hoḍe ka:nu* means, (for instance) the *doḍḍa sampige* stream flows through the middle.)

(BG) *A: kaḍeka guḍde, i: kaḍeka uwe guḍḍe.*
(There's a hill on that side, there's a hill on this side.)

(BG) *I: eraḍu guḍde se:ri hoḍe a:gottu… hoḍe ka:ninalli hoḍe bidiru uwe ade.*
(Put the two hills together, and you get a *hoḍe*… you find *hoḍe* bamboo growing in a *hoḍe ka:nu*.)

5.3.10 *Saṇṇa ka:nu*

A habitat type peculiar to the southern Western Ghats, where the Solega live, is the much-celebrated 'shola forest'. These are small patches of evergreen forest (often described in the conservation literature as 'stunted') that are found on high hill- and mountaintops, and are surrounded by large expanses of treeless grassland. These directly contrast with the *doḍḍa ka:nu* described above in that they form isolated clumps of trees, whereas the *doḍḍa ka:nu* is regarded as a continuous stretch of evergreen forest.

Extract XX (S)

(MRM) *I:ga na:vu beṭṭa:da me:le i: saṇṇa saṇṇa ka:nu irtavalla?*
(So there are these numerous small evergreen forests on the tops of mountains, right?)

(BG) *A:kaḍe ade se:ṭa ka:nu. A: se:ṭa ka:nu ishṭe: a:ga:la:ka ade. Enduru, adu ondu keḷasari kaḍeli ade, eḍage ka:nu.*
(There's one over there—*Se:ṭa ka:nu*. That *Se:ṭa ka:nu* is this small! And then, on one side of that, there's the *Eḍage ka:nu*.)[9]

…

…

(MRM) *Beṭṭa me:le iddale saṇṇa ka:nu.*
(So if it's on top of a mountain, it's a *saṇṇa ka:nu*.)

(AS) *Pa:se tumba irutte, allava? Mara pa:se…*
(There's a lot of moss/lichen, isn't there? Tree moss…)

(MRM) *Ã, mara pa:se adella? Kappu pa:se.*
(Yes, there's tree moss, isn't there? Black moss.)

(BG) *Ã, bonte bonteve! Adu ishṭu ishṭu udda neḍdide keḷeka. Marada kombu me:le bandu hi:ge… a: pa:seya me:le phalli pera:ṇi ha:vu sette bodde ella baḷte ade.*
(Yes, heaps of it! It hangs down this far! Like this, from the branches of trees… animals like geckos, snakes, green snakes etc. all live on the moss.)

The abundance of moss in this forest type is quite striking, and immediately apparent to even a casual observer. Shola forests also have a distinctive suite of tree species growing in them, even though they are found at the same altitude as the evergreen forests (Extract XXI)

Extract XXI

(JS) *Male ka:ḍina maragaḷu be:re, saṇṇa ka:ḍina maragaḷu be:re.*
(The trees in the evergreen forest are different from the trees in the shola forest)

5.3.11 *Oḍḍuga:ḍu/kallo ḍḍu*

The name of this landscape type suggests that it is visually dominated by piles (*oḍḍu*) of rocks (*kallu*). Although *odduga:ḍu* can be a good source of certain kinds of honey, such places tend to be the natural habitat of dangerous wild animals, such as bears. It is generally advisable not to wander into such places by oneself.

[9] BG goes on to relate the myth of how the *Se:ṭa ka:nu* came into being.

Extract XXII

(BG) *Oḍḍuga:ḍu. Kalloḍḍu. I:ga aḍkugallu, matte e:ru kallu, etti maḍagida oḍḍu, i: taradalli ade: adu. Adu oḍḍuga:ḍu. Kalloḍḍu, ba:ri ba:ri tad-dira oḍḍu. Alli karaḍi ue ade, a: ja:ga:dalli huli ue va:sa ma:ḍa:de. "Oḍḍi.ka ho:ga:daka a:ga:dille! karaḍi ade!" A: oḍḍi.ga ja:ṭa:de. Oḍḍu iddale karaḍi a:ru uḍtade e:na ma:ḍde? Negedu uṭṭattu nanaga ondu dina...*[10]

(*Oḍḍuga:ḍu, kalloḍḍu.* Consider *Aḍkugallu*, and *E:ru kallu*, the rocks in these places are piled up; that's what an *oḍḍuga:ḍu* or *kalloḍḍu* is like. Lots of rocks piled on top of one another. There are bears there, tigers also live in those places. "You shouldn't go to the *oḍḍu*! There are bears there!" (people would say.) The *oḍḍu* is off limits. When I was in the *oḍḍu*, a bear attacked me—what did I do? I was attacked one day...)

(MRM) *Maragaḷu i:te alli?*

(Are there trees there?)

(BG) *Doḍḍa mara ille.*

(There are no big trees.)

(MRM) *E:n e:na mara beḷeda:du alli? Oḍḍuga:ḍinalli?*

(What trees grow there? In the *oḍḍuga:ḍu*?)

(BG) *Adu ashṭe ide marave. Icci mara, kallicci mara, biḷicci, handicci, go:ṇi, a:mele maruḷi... a: maruḷi hu:vina je:nu baṇḍa:de! Pairu je:nu...*

(Just the same ones (you have here). *icci* (*Ficus nervosa*), *kallicci* (*Ficus tinctoria*), *biḷicci* (*Ficus amplissima*), *handicci* (?), *go:ṇi* (various), and then *maruḷi* (*Indigofera* sp.?)... honey from the *maruḷi* flower is delicious! It's good honey...)

The bear attack story related by BG contains many elements typical to an *oḍḍuga:ḍu*—the presence of animals like bears and porcupines, large boulders, sheer drops, plants characteristic of this kind of landscape, and finally, the presence of honey, which brings people to places like this.

5.3.12 *Bo:ḷiga:ḍu/beṭṭa bo:ḷi*

Grassland found on the top of mountains is referred to as *beṭṭa bo:ḷi*. Such places are devoid of trees (or indeed, anything other than grass or other low-growing herbs), and are therefore appropriately named *bo:ḷi* ('shaved head' in Kannada).

[10] BG goes on to tell of an incident in which he was attacked by a bear while looking for honey in an *oḍḍuga:ḍu*.

Extract XXIII (S)

(MRM) *I:ga na:vu beṭṭa me:le ho:da:ga bo:ḷi ga:ḍu endivẽ.*
(Now, when we go up a hill, we call that *bo:ḷiga:ḍu*.)

(BG) *Ã, beṭṭa bo:ḷi, mara ille alli. Bo:ḷiga:ḍu endale alli biri hullu ashṭe ta:ne ira:du, adakka bo:ḷi.*
(Yes, *beṭṭa bo:ḷi*, there are no trees there. *Bo:ḷiga:ḍu* means there's only grass, that's why it's '*bo:ḷi*'.)

(MRM) *Alli beḷeanta hullu e:n e:na beḷeda:du?*
(What are the grasses that grow there?)

(BG) *Alli beṭṭa na:le hulle: beḷedade, ba:ṇe hullu beḷeda:de, handi ba:ṇe hullu beḷeda:de, a:ga matte, i: uṇuga:na hullu beḷeda:de, innue oḍḍa:na hullu beḷedade, matte ko:ḷi hullu, matte navaṇe hullu, ashṭella ue alli irtade adu. A: ja:ga:dalli innue ambu hullu beḷeda:de, intadu innue gareke hullu, adella ue banda:de. A:mele oḍḍoḍḍu barebare me:leve ond –ondu bare sã:ya ade summa:ne illi huṭṭitu endale summa:ne abburu oitade adu.*
(*Beṭṭa na:le* grass (?) grows there, *ba:ṇe* grass (?), etc… all these are found there. *Ambu* grass (?) and *gareke* grass (?) also grow in that place. And then, you find clumps of *bare sã:ya* (rock fern) on rocks and boulders. They just grow there and cover everything.)

(MRM) *Adu oḍḍinalli ira:du?*
(They grow on the piles of rock?)

(BG) *Oḍḍoḍḍinalli, bare bare me:le. Ashṭella adu oḍḍinalli irtade.*
(On the piles of rock, on the boulders. They all grow on the piles of rock.)

(BG) *...Mara sikka:dille alli. Maḷe bandaluve tappisa:dakka a:ga:dille, ella angaṭṭi keḷeka ue me:leka ue. Adakka:gi bo:ḷi beṭṭa:ka na:vu e:ra:dille.*
(… You don't find trees there. You can't take shelter when it rains, it's all open above and below. That's why we don't climb up to the *bo:ḷi beṭṭa*.)

(BG) *… E:nendale, ho:daluve, ba:ṇe hullu nugguru o:ga:dakka a:ga:dille. Ha:geve bagedu, nittu uḍtu jardru ho:ga:de kashṭa. Ba:ṇe hulliga teṇṭe ipipa:ṭi ittu.*
(… So, even if you go (to the *beṭṭa bo:ḷi*), you can't go through the *ba:ṇe* grass. It's difficult, because you have to lean forwards and push through. The clumps of *ba:ṇe* grass used to be this big!)

In the above exchange, both BG and MRM observe that *bare sã:ya* 'rock fern' are plentiful around piles of rocks '*oḍḍu*' and boulders '*bare*', within the context of describing a *beṭṭa bo:ḷi* 'hill grassland'. This is unsurprising, as rocks and boulders frequently occur in otherwise featureless grasslands. *Oḍḍu*, however, should not be confused with *oḍḍuga:ḍu* (described above), as the former refers to individual rock formations, while the latter describes a region dominated by boulders and piles of rock.

BG's statement that "we don't climb up to the *bo:ḷi beṭṭa*" also needs to be understood in context. Open grasslands may well be avoided during periods of rain due to a lack of shelter, but it seems unlikely that they would be shunned altogether. After all, these grasslands are important sources of various grass species, many of which, especially *ba:ṇe* grass, are used by the Solega for the construction of dwellings.

5.3.13 *Guḍḍega:ḍu, bo:rega:ḍu*

These terms refer to another kind of vegetation associated with elevated areas, namely hill- and hill-slope forests. This habitat can be home to a variety of mammal species during the rainy season. During the dry season, however, when the water sources in the hills start to disappear, animals such as elephants, sambhar, buffalo, mouse deer and barking deer prefer to remain in the flatland forests (*dimba ka:ḍu*, *maṭṭaga:ḍu* or *gaddega:ḍu*), where some streams and water holes still contain water. This time also coincides with the main yam harvesting season, and young Solega are strongly advised by their elders to only go to the hill forests to dig for yams. This is primarily a strategy to avoid running into thirsty (and therefore probably irritable) elephants that would be wandering the flatland forests during this time.

Extract XXIV

(MRM) *Bo:rega:ḍu andare guḍḍe, guḍḍega:ḍu. bo:re gaṭṭu anta. Na:vu he:ḷdivẽlla, adu ni:vu hindina ka:lada de:no he:ḷtiddiralla?*
(*Bo:rega:ḍu* means 'hills', hill forests. We call them *bo:regaṭṭu*, don't we, what were you saying about how it was in the old days?)

(BG) *Ã, bo:re gaṭṭu. "Oḍa:ginalli no:ḍru ba:ri bo:re gaṭṭudikka! A:neginega bandu uḍtave, gaddega:ḍiga biḍa be:ḍi!" enduru he:ḷdaru.*
(Yes, *bo:regaṭṭu*. "Go have a look at the *bo:regaṭṭu* on the mountain slopes! You'll run into elephants, don't go to the flatland forest!" they would say.

(MRM) *Ni:vu ka:ḍiga oiṭiddiru, o:darue idu... "bo:regaṭṭinalli ku:tu geṇasu a:gedumo, maṭṭa:ga:ḍinalli a:ne ade."*
(You were going to the forest, and then... "Stay in the *bo:regaṭṭu* and dig for yams, there are elephants in the flatland forest."

(BG) *Gadde ga:ḍiga o:dale a:negave, tappisa:dakka:ga:dille... gaddega:ḍina iradille, beṭṭa:da hoḍa:ginave iddu a:mele banduma:du. Hoḍagu pakka, me:leka beṭṭa, a:ne solpa e:rakka:ga:dille.*
(If you go to the flatland forest, you'll only find elephants, you can't avoid them... you don't stay in the flatland forest, you go halfway up the mountain, and later you should come back. Close to halfway—the mountain rises above, the elephants don't really climb up.)

(MRM) *A:ga guḍḍega:ḍina a:ne idda:ga maṭṭa:ga:ḍu iḍdurtiddiru*
(So when the elephants are in the hill forests, you'd go down to the flatland forest?)

(BG) *Maṭṭa:ga:ḍina a:ga:de ille, ella ṭa:im uwe maṭṭa ga:ḍina a:ne idda:geve. Bo:reve ira be:ku. Alli ondu ga:ḷige geṇasa aggeduma:du. Maṭṭa:ga:ḍinalli ishṭu geṇasina ambu iddaluve aggetaralle.*
(You just don't go to the flatland forest, there are always elephants there. You should just stay on the *bo:re*, and dig for yams there for a while. You don't go to the flatland forest even though there are so many yams there.)

…

…

(BG) *Illi keḷaga bellatta murṭiyalli KKMna heḍati, solpa varsadalliye adu. Ave:na ma:ḍtidda, maṭṭaga:ḍinalli geṇasina guḷiya agetidda…*[11]
(Down here in<PLACE NAME>the wife of KKM, she was only young. What was she doing, she was digging for yams in the flatland forest…)

In an unrecorded informal interview, MRM mentioned that these forests tend to have a mix of big and small trees, of which the prominent species were *honne* (*Pterocarpus marsupium*), *matti* (*Terminalia crenulata*), *ta:re* (*Terminalia bellirica*) and *ne:raḷe* (*Syzygium cumini*). Rocks and grass are also commonly found on *guḍḍega:ḍu*. This information is confirmed by JS in the first elicitation session:

Extract XXV

(JS) *doḍḍa ka:ḍu adu, mara miks irutte. matte baṇḍe solpa irutte*
(It's a big forest, there's a variety of tree species. You also find a few boulders)

5.3.14 *Dimba ka:ḍu*

In the preceding discussion, it became apparent that within the Solega-dominated biogeographical area, the distinction between hill/mountain forest and flatland forest is a highly salient one, not only because of the obvious physical differences between the two forest types, but also as a result of the very real dangers posed by elephants in the latter habitat. As mentioned previously (see Extract XI and the subsequent discussion), flatland forests (*gaddega:ḍu* or *maṭṭa:ga:ḍu*) in the Solega experience can occur both at elevation, and in the lowlands. While the low flatland forests form part of the *orrega:ḍu/kutarega:ḍu/na:ḍu ka:ḍu* macrohabitat, which lies outside the traditional lands of the Solega, the flatland forests at elevation are very much within the *male ka:ḍu/tho:pu ka:ḍu/ka:nu ka:ḍu* macrohabitat inhabited and utilised by the Solega.

In the following two extracts, all speakers agree that *dimba ka:ḍu* indicates flatland forests that occur at high altitudes. Some of the speakers were in fact residents of the nearby *po:ḍu* 'village' called *Keredimba* (*kere* 'reservoir'+*dimba*), named simply for being situated on a plot of flat land next to a reservoir. Such *dimba* areas are preferred by the Solega for habitation and agriculture, although they have the disadvantage of being frequented by elephants in search of water in the dry season (see Sect. 5.5).

[11] BG goes on to tell the story of the wife of an acquaintance who was trampled to death by an elephant while collecting yams in a flatland forest.

Extract XXVI

(MRM) *Dimba endivẽlla?*
(We say *dimba*, don't we?)

(BG) *I:ga idu me:le malaki beṭṭa, adu bo:ḷi, keḷage—malaki beṭṭa:d.inda itta:ka dimba. e:vada—kancaḷa dimba. Me:leka beṭṭa, keḷeka kancaḷa dimba allava?*
(So, above we have Malaki mountain, with grassland, and below—this side of Malaki mountain is a *dimba*. Which one—*Kancaḷa dimba*. Malaki mountain, and below, *Kancaḷa dimba*.

(MRM) *I: maṭṭa:ga:ḍiga dimba.*
(The flatland forest is called a *dimba*.)

(JY) *Beṭṭa me:le a:gi, keḷage ondu ma:ḷa ada:darinda adu dimba.*
(When there's a mountain above, and an open forest below, that's a *dimba*.)

(MRM) *Ã. Idu gaṇṭe dimba, matti arre dimba, e:role dimba, allinda bandu illi...*
(That's right. *Gaṇṭe dimba*, *matti arre dimba*, *e:role dimba*, and if you come this way…)

(BG) *E:ru kallu dimba. A: tara. A:mele, idu soreka:ye beṭṭa ittaka, du:managadde. Adu gaddeve.*
(*Eru kallu dimba*. Like that. And then, this side of *Soreka:ye* mountain, there's *Du:managadde*. That's also just flat marshy land.)

…

…

(MRM) *Saṇṇa guḍdeyinda keḷaka iddale adu maṭṭaga:ḍu... dimba ka:ḍu alliga. Maṭṭa pu:ra banda:della, beṭṭa:dinda aḍi:li, adakkella dimba ka:ḍu.*
(There's flatland forest when you go down small hills… that's called *dimba ka:ḍu*. It's totally flat, right, at the base of a mountain, all that's called *dimba ka:ḍu*.)

(BG) *Ã dimba ka:ḍu. A: dimba ka:ḍinalli, e:na—tho:pu.*
(Yes, *dimba ka:ḍu*. And that *dimba ka:ḍu* becomes—*tho:pu*.)

(MRM) *Avella se:rdale tho:pu ka:ḍu.*
(When it all joins up, it's a *tho:pu ka:ḍu*.)

In the previous extract, BG and MRM were joined by JY, who joined the conversation briefly to explain a couple of lexical items. In the following extract, the young men agree with BG's definition of *dimba*. JS goes on to contrast flat mountain- and hilltops with pointed ones.

Extract XXVII

(NG) *Dimba andare maṭṭa.*
(*Dimba* means 'flat'.)

(JS) *Pla:ṭ irutte. I:ga beṭṭa irutte solpa... solpa slo:pa:gi uṭṭu pla:ṭ irutte, no:ḍi.*
(It's flat. Now, there's a mountain, it slopes (up) a bit, and becomes flat, see?)

(NG) *Adakke daṭṭa ka:ḍu, daṭṭa ante:ḷi andare sama:na:va:gi.*
(That's a *daṭṭa ka:ḍu*, *daṭṭa* means 'level')

(JV) *Dimba ka:ḍu andare beṭṭa:da me:le solpa maṭṭa irutte, andare onde leval irutte, beṭṭa:da me:lgaḍe. Adanna dimba.*
(*Dimba ka:ḍu* means it's a bit flat on top of a mountain, meaning it's all on the same level on the mountain. That's a *dimba.*)

(JS) *Adu cu:pa:gi iddare, adu bo:ḷi anta baratte, tumba maṭṭa andare, adu dimba.*
(If it's pointy, we call that a *bo:ḷi*, if it's very flat, that's a *dimba.*)

5.4 Nomenclatural Considerations

The preceding discussion indicates that the morpheme *ka:ḍu* is highly polysemous, and can be used in words that indicate landscape types and forest/vegetation types, which are recognized on the basis of biological, topographic and geological features, as well as human land use patterns. Some types of *ka:ḍu* are named on the basis of more than one of the above features, as in the case of *kutarega:ḍu*, where the fact that this is a lowland forest is as salient as the fact that only small trees grow here.

Table 5.3 presents a list of all Solega words that contain the morpheme *ka:ḍu*, along with the primary features that speakers mentioned in the above extracts as being diagnostic of a particular type of *ka:ḍu*. The *ka:ḍu* terms are divided into two categories, 'macro' and 'local', which is a reference to speakers' assertions that many instances of the latter type can often be found in a region labeled with one of the former. For instance, *doḍḍa ka:nu* 'evergreen forest', *saṇṇa ka:nu* 'shola forest' and *bo:ḷiga:ḍu* 'hilltop grassland' can all be found within the realm of *beṭṭaga:ḍu* 'mountain forest' (Extract V). This distinction between 'macro' and 'local' is not a clear-cut one, however, as some terms from the latter category are also said to include other, 'local' terms in a partonymic sort of relationship. One such example is *(doḍḍa) ka:nu ka:ḍu* 'evergreen forest', which is a large expanse of forest, and can extend over river valleys, mountain slopes and elevated stretches of flat land. The corresponding *aḷḷaga:ḍu* 'river forest', *bo:re ka:ḍu* 'slope forest' and *dimba ka:ḍu* 'flatland forest' would then be considered parts of the larger *ka:nu ka:ḍu*. One generalization that could be made about the two categories of *ka:ḍu* is that the 'macro' terms can be said to contain a variety of vegetation types, while even the larger, more inclusive basic terms (such as *doḍḍa ka:nu*) consist of only one dominant vegetation type, i.e. large trees.

The possible partonymic relations between the various types of *ka:ḍu* are therefore determined by commonsense notions of topography and vegetation (there obviously cannot be a *kutarega:ḍu* 'lowland forest' within a *beṭṭa ka:ḍu* 'mountain forest', just as there cannot be a *bo:ḷiga:ḍu* 'hilltop grassland' inside an *aḷḷaga:ḍu* 'river forest'), and do not represent a strict hierarchy of fixed ranks or exclusive relations of set inclusion. 'Macro' and 'local' are used here only for analytical convenience, with the former denoting broad biogeographical domains, and the latter representing more localized macrohabitats; I am not suggesting that there is a sharp language- or cognition-based ontological distinction between the two. Still, it is interesting to note that the 'macro' terms given in Table 5.3 all focus on the *elevation* feature, while the 'local' terms have a much bigger range of primary features (with *elevation* and *topography* figuring prominently in a majority of cases). This is hardly surprising, since three of the

Table 5.3 Solega *ka:ḍu* terms and the primary features used to define them (as determined from the transcripts presented in the text)

Ka:ḍu type	English gloss	Primary defining feature
'Macro' terms		
male ka:ḍu	Highland forest	Elevation
tho:pu ka:ḍu	Highland forest	Elevation
na:ḍu ka:ḍu	Plains vegetation	Elevation
beṭṭaga:ḍu	Mountain forest	Elevation
maṭṭaga:ḍu	Flatland forest	Elevation/topography/wildlife
'Local' terms		
oḍḍuga:ḍu	Boulder field	Geology
eḷa:vuga:ḍu	Early fallow	Human action (agric.)
oṭṭuga:ḍu	Late fallow	Human action (agric.)/ vegetation
benda ka:ḍu	Ash bed	Human action (fire)
aḷḷa ka:ḍu	Riverine forest	Hydrology
koḷḷa ka:ḍu	Valley forest	Topography
bo:rega:ḍu	Slope forest	Topography
hoḍe ka:nu ka:ḍu	Gorge forest	Topography
guḍḍega:ḍu	Hill forest	Elevation/topography
dimba ka:ḍu	Flatland forest	Elevation
saṇṇa ka:nu ka:ḍu	Shola forest	Elevation/vegetation
bo:ḷiga:ḍu	Hilltop grassland	Elevation/topography/vegetation
orrega:ḍu	Foothills vegetation	Elevation/vegetation
kutarega:ḍu	Plains vegetation	Elevation/vegetation
begga:ḍu	Waste forest	Lack of utility
doḍḍa ka:nu ka:ḍu	Evergreen forest	Vegetation/climate

'macro' terms (*male*, *tho:pu* and *na:ḍu*) overlap with the highland/lowland distinction that is so salient to considerations of Solega identity. The one exception in this respect is *maṭṭaga:ḍu* 'flatland forest', which also has a *wildlife* feature associated with it. Again, this is to be expected, for the danger posed by elephants in such locations (Extract XXIV) is something the Solega are constantly aware of.

5.5 Landscape/Forest Terms and Seasonality

In many of the landscape/forest term explanations presented above, there seems to be a strong element of seasonality which, to an extent, shapes Solega perceptions of these landscapes, and determines the time and manner in which the Solega may utilize the resources contained therein. The terms *guḍḍega:ḍu* 'hill forest' and *maṭṭa:ga:ḍu* 'flatland forest' are, for instance, frequently contrasted, as animals such as elephants are more likely to be found in the latter landscape type during the dry season. As a result, the Solega may only safely gather yams in the hill forests during this time, even though this food source is also quite plentiful in the flatland forest.

The dominant forest type of the region, the *doḍḍa ka:nu* 'evergreen forest', also displays seasonal characteristics, in spite of the impression of changelessness that the English gloss 'evergreen' tends to impart. Even though it is cool and dark year round, and has a permanent water supply, it is home to animals like the buffalo and pangolin for only part of the year. These animals prefer to live in the lowland forests during the rainy season. Honeybees, on the other hand, start to arrive in the *doḍḍa ka:nu* in March–April, just prior to the start of the rainy season, when the flowers of the *honne* tree start to bloom (see Chap. 6 for more details on seasonal cycles). Finally, certain types of *aṇabe* 'mushrooms' can be found in the *doḍḍa ka:nu* only in the rainy season—these include *doḍḍaṇabe*, *ko:ḷi aṇabe*, *beṇḍe aṇabe* and *eṇṇe aṇabe*, which start to appear in June–July.

While the examples given above are testimony to the remarkably detailed knowledge that the Solega have of forest resource availability and the movements of wild birds and animals, they are not nearly as impressive as my consultants' ability to accurately report the phenological status, at any given moment, of any of the 300 or so tree species known to them (Table 5.1 and Fig. 5.5). This awareness of tree life cycles is inextricably linked with the knowledge of the kinds of habitats in which they normally grow. Many of the landscape/forest terms discussed above are strongly associated with very specific suites of tree species (which may or may not overlap with the species growing in other forest types). By combining such distributional information with the phenological information for practically every single tree species known to them, the Solega have the potential to construct a detailed and dynamic three-dimensional botanical map of their ancestral lands. When other information, such as their knowledge of animal movements and water availability, is overlaid onto this map, it results in a map of accessible resources that is constantly updated with the passing of the months. Figure 5.4 represents a much simplified version of this concept, with the time dimension depicted along the x-axis, and three forest types arbitrarily arranged along the y-axis (the complex spatial relations between the three landscape/forest types have been omitted for the sake of clarity). In the month of March, for instance, the *ka:nu ka:ḍu* would still be inhabited by buffaloes, civets, monkeys, etc., and honeybees would have only just started to arrive from the lowland forests, to coincide with the flowering of the *honne* tree. Trees like the *kara:va:di* (*Persea macrantha*) would already be producing ripe, edible fruit, while others, like the *matti* and *bejja*, would still be bare. At the same time, in the *guddega:ḍu* and the *maṭṭa:ga:ḍu*, one would expect to find the animals and plants mentioned earlier, and react accordingly to efficiently and safely collect the appropriate resources. The Solega therefore have the potential to perceive their ancestral lands as a patchwork of forest types, each one changing with the seasons to yield different resources at different times of the year.

Figure 5.5 shows four landscape/forest types arranged in a single linear transect. The real ecosystem is, of course, not as simple as depicted, because the various landscape/forest types would form a two-dimensional mosaic with far more complicated boundaries, overlaid on the third dimension of elevation. Still, the figure provides a good overview of the kinds of phenomena that Solega people would associate with each location, in different seasons.

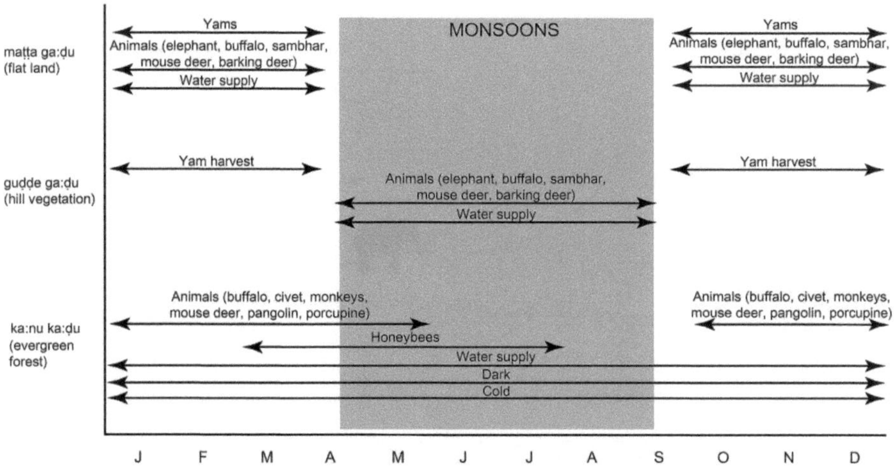

Fig. 5.4 The changing resources to be found in three landscape/forest types over an annual cycle. Previously published in [8]

5.6 Patterns of Usage

A cognitive map such as the one depicted in Fig. 5.5 would allow Solega people to identify specific locations in their ecosystem at specific times of the year, with the aim of either exploiting such locations for natural resources, or avoiding them for fear of dangerous animals. In the preceding paragraph, I was careful to say that such uses of the cognitive map (if, indeed, such a representation really does exist) are a potential feature of Solega cognition. Potential, for two reasons: the first is that as Kitchin [64] has pointed out, the word 'map' may in itself be misleading, as it implies the existence of a complete cartographic map in the mind, all the components of which are activated at all times, waiting to be 'consulted' by the mind's eye. In the case of the dynamic Solega cognitive map, a stack of cartographic maps, one for each part of the seasonal cycle, would be required. A far more likely scenario is that Solega people 'construct' relevant parts of the map, as and when they are required (i.e., when planning a trip), from their store of encyclopaedic knowledge associated with particular locations for particular times of the year. I elaborate on this point in the following paragraphs. The second reason is that in spite of the impressive lexical diversification in landscape terminology, my corpus contains surprisingly few tokens of the different landscape and forest terms described in this chapter. The words *ka:ḍu* and *na:ḍu* were used very frequently—the former to refer to the forest in general, and contrasting with places of human habitation, or the deforested plains region dominated by non-Solega; the later term is used to indicate all lands inhabited by non-Solega. A common feature of the very few instances where a more specific landscape or forest term was used was the need to indicate that the entity or phenomenon being talked about was commonly found in, or restricted in its occurrence to, that particular forest or landscape type. For example:

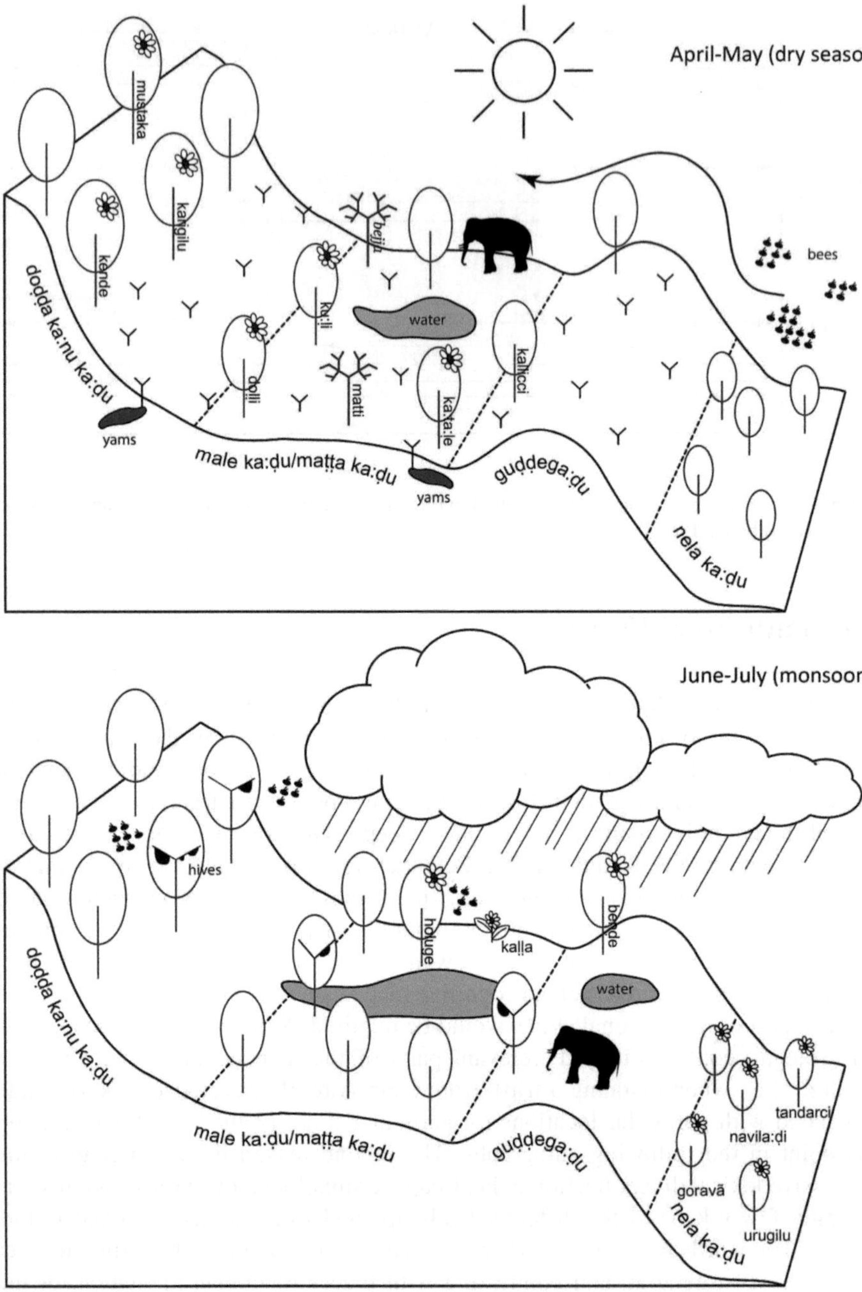

Fig. 5.5 The phenological signatures of four landscape/forest types at two different times of the year

Ka:nina oḷage binna buḍu ade
There are fireflies inside the evergreen forest.

I: pa:riva:ḷa doḍḍa doḍḍa bareli, oḍḍinave ira:du
This pigeon is only found on big rocks, in the boulder field.

It is highly probable that even though landscape and forest terms are used frequently in everyday Solega speech, I was unable to document such utterances due to the very specific contexts in which they are spoken. Naturally, many of my interviews, where I ask specific questions about the habitats of various plants and animals, contain these words, but even in lengthy recordings where one or more speakers discuss forest-related matters, the preferred strategy is to use the general term *ka:ḍu*. Two different sources of information suggest that the use of landscape/forest terms is more widespread than is attested in the 'naturalistic speech' component of my corpus. The first is, ironically, the interview data I have presented in this chapter. BG, the older consultant interviewed here, frequently used direct quotations (uttered by real or fictional characters) in his narratives to create brief conversations, and provide a sense of immediacy to his stories; examples of this strategy can be clearly seen in Extracts XII, XXII and XXIV. In the latter two extracts, which presumably represent an older adult admonishing a younger person, the landscape terms are explicitly mentioned as places to go to, or avoid. This suggests that these words are used in the context of making travel plans, or at the very least, asking the hearer to avoid or go to a particular location. I was able to locate two other examples of landscape term usage, which support this hypothesis:

Matte neḍi matte! I: oḍḍu ka:ḍiga ho:gõ!
Come on, start walking! Let's go to the boulder field!

A: hoṭṭinalli ho:gi turuguruve mudiya no:ḍru...
We went to the fallow land to forage and look for new hives...

Incidentally, the above utterances were spoken by BG before the interview on landscape terms had even begun; he simply volunteered a story about his past experiences with farming and foraging soon after we had sat down to talk, and so these two utterances, offered in quick succession, represent good examples of the use of landscape terms in naturalistic speech.

The second type of data that can be mobilised to make a case for the frequent use of landscape/forest terms is the large set of toponyms (discussed in detail in Sect. 3.7) which contain, as the last part of the name, the identity of the landscape or hydrological feature on which the toponym is based. Once again, given that these landscape and hydrological terms are part of place names, one would expect them to be used frequently in the context of travel, as in:

Alli serkae beṭṭa ho:gi hola eḍiya:du.
Then we went to *Serkae* Mountain, and cultivated some land.

This coupling of landscape/forest terms with ideas of motion or travel meshes well with Gärling et al.'s [194] claim that "*information about the environment (i.e. the cognitive map) is acquired in connection with the formation and execution of travel plans*" (p. 22). Although their theory appears to have been developed primarily for urban environments, many of its features appear directly applicable to the Solega situation. Accordingly:

> People form travel plans with the general goal of reaching desired destinations without having to invest too much effort… To attain this goal, information about important properties of the environment must be gained. These properties include the functions of places (e.g. whether there is a shop, what kind of shop it is, and so forth), the attractiveness of place (if a place is, for instance, a shop, whether the desired goods are likely to be available, and so forth), the identity of places (i.e. the characteristics, name and/or perceptual characteristics that distinguish a particular place from all other places), where the places are located, and how one can travel to the places. (p. 20)

This model provides a simple, utilitarian rationale for why people might categorise landscapes or forests in particular ways, and why the names for these categories might only occur in particular contexts in Solega. Solega people categorise their environment in terms of their interactions with particular places—these might include seeking out places with resources, avoiding dangerous places, and not lingering in places with no utility. Naturally, such considerations are central to the potential travel plans that Solega people might have to make on a daily basis. "*Frequency of use*" (Evans 2003) and "*customary ways of behaving which confirm and reinforce* [the status of named concepts] *as social institutions*" [200] have been identified as important ways in which culture can affect language structure. The Solega landscape terms, with their strong connection to cultural activity, appear to be significantly affected by these factors.

5.7 Conclusion

The Solega words that denote forest or vegetation types also form an integral part of the landscape lexicon, mainly because the forest is such a conspicuous part of the Solega's visual environment. As shown above, the word *ka:ḍu* is highly polysemous, and combines with many of the geological landscape terms to denote places that can be glossed as either physical landscapes, forest types, parts of a forest, or even areas of human activity. The Solega *ka:ḍu* words carve up the entirety of their ancestral lands into discrete units, which appear to contrast with each other largely on the basis of topographical location, the suites of plant and animal species to be found in each region, major human activities associated with each location and prominent abiotic features of the landscape.

As the average Solega has internalized the phenological profiles of all the major flowering plants named in his/her language, and is also aware of the migratory patterns of various organisms such as honeybees and elephants, s/he is able to 'know' what a given patch of forest is 'doing' at any given time. The great importance, of the living elements of the landscape is also reflected in the fact that about half of all Solega place names make reference to a plant or animal species that is associated with a particular locality. As a result, the Solega likely perceive their natural environment as a mosaic of discrete, but interconnected, patches of habitat comprised of both abiotic landforms and biological organisms, the condition of which is constantly updated in sync with the cycle of the seasons.

Chapter 6
Signs and Relationships

6.1 Introduction

Having examined, in some detail, the physical and biological environment that the Solega live in, we now turn our attention to the realm of the intangible—a realm populated by multitudes of meaningful signs and relationships which enable Solega people to navigate a world that is at once familiar, complex and sometimes dangerous. Knowledge of the existence of a particular type of flowering plant in the evergreen forest is no doubt important to a Solega, but equally important are observations that the plant grows in a particular habitat, flowers at a particular time of the year, is eaten or visited by a particular type of animal or insect, and can be used by humans for a particular purpose. After all:

> Organisms are not separate; they are linked. They are linked by their own relations—sign relations. These relations themselves should be our objects [of study] if we want to understand the life process of living communities, life as it happens in ecosystems. From a semiotic point of view, ecological communities are not sets of organisms (or species) as elements; instead, a community is a composition of relations between the organisms or species [201].

The principal intent of the author of the above quote (and of other 'biosemioticians') is the "semiotic description of biological communities", rather than the elucidation of the ways in which "sign relations" are encoded in human languages. Indeed, some biosemioticians are eager to distance themselves from the so-called "glottocentrism that is enshrined in cultural analysis" [202], i.e. the efforts of linguists and anthropologists, who concentrate on verbal and artefact-based symbols and signs, at the expense of natural ones. Nevertheless, many of the central concepts and questions of the biosemiotic enterprise can be easily transferred to the field of ethnobiological enquiry. The idea of an organism's Umwelt, its subjective world, was first proposed by Jakob von Uexküll, and later redefined [203] as a "personal semiotic space". A 'semiosphere', then, is "the set of all interconnected umwelten", which is quite different from the physical environment in the scientific sense. In the

A. Si, *The Traditional Ecological Knowledge of the Solega*, Ethnobiology,
DOI 10.1007/978-3-319-24681-9_6

case of humans, it could be argued that the signs that make up an individual's Umwelt are often referred to explicitly in that individual's speech. By investigating the utterances of a linguistic consultant, then, an ethnobiologist can begin to build an accurate picture of the consultant's personal semiotic space, and achieve a more sophisticated appreciation of his/her perceptions of other organisms. In the following sections, I employ an unashamedly anthropocentric, glottocentric approach to Umweltforschung. In this chapter, I also make frequent reference to (1) the non-verbal signs that Solega speakers perceive in their daily interactions with other organisms, and (2) the non-verbal interactions that Solega speakers perceive to be occurring between other, non-human organisms. It is hoped that such an approach will go some way towards describing the perceived semiosphere of the Solega—this would cover not only the signs that an individual is aware of, such as the call of a particular species of bird that indicates a meteorological event, but also those that do not quite enter the individual's consciousness, such as the various cross-modal cues that are required to make a positive identification of a plant or bird in its natural habitat.

The correct identification of plants and animals plays a key role in virtually all aspects of Solega life: certain leaves are said to be 'pleasing' to the Solega gods or *de:varu*, and are preferred for the construction of shrines (*sappara*); different species of honeybee need to be distinguished so that the most appropriate strategies and locations for honey collection can be chosen; and all manner of forest trees are readily identified for the various edible and non-edible natural products that each regularly provides. However, this knowledge of 'how things are named' and of 'what use can be derived from each named entity' represents but a small proportion of the average Solega speaker's total understanding of the natural world within which s/he is situated. Just as important as the naming of things is, for him/her, an awareness of the interactions between organisms and (what Western science would call) the physical environment, between different organisms, between humans and the physical environment, and between humans and other organisms.

When asked to identify the plants growing along a forest path, Solega consultants almost always provided one of three different types of responses:

"That is an X. You use its [part] for [function]."
"That is a Y. It is the favourite food of [animal name]."
"That is a Z. It has no use."

Statements such as these demonstrate that most organisms are perceived in terms of their relationship(s) with other organisms, including humans, and that such relationships can be almost as salient (if not equally salient) as the separate identities—X, Y and Z in the examples used above—of the organisms being talked about. Both the identification of an organism, as well as the recognition of the relationships it might have with other organisms, require the reception and correct interpretation of a host of signs emanating from an organism, an observed interaction, or both. In some cases, a sign emanating from an organism might be indicative of an unrelated biological or physical phenomenon, such as the call of a bird heralding sunset. While these signs are invariably non-linguistic, they can be, and are, talked about in Solega, with the result that many of the phenomena indicated by such signs form an explicit part of the Solega ethnobiological knowledge system. An investigation of the ways in which Solega speakers encode knowledge of such signs and phenomena in everyday language should yield valuable insights into their perceptions of the natural environment. These insights should also help construct a far more detailed picture of Solega perceptions than that obtained from a consideration of their ethno-classification system alone.

While this chapter is not meant as a formal semiotic analysis of the Umwelt of the Solega speaker or of the other Umwelten that share a semiosphere with him, it nevertheless acknowledges the conviction held by many biosemioticians that "the scope of semiotics encompasses the whole of the *oikoumene*, the entirety of our planetary biosphere," and that semiosis "must be recognized as a pervasive fact of nature as well as of culture" (Sebeok 1977 in [204]). Members of traditional societies are often credited with possessing an "impressive" knowledge of their habitat [9], and I argue below that it is actually the correct interpretation of a host of non-verbal signs that makes this knowledge accessible. The Solega speaker's knowledge of nature[1] thus becomes a matter of intercepting signs and interpreting their significance.

6.2 Types of Signs

As mentioned earlier, Solega people, as presumably do all humans, draw on the vast amount of information encoded in environmental signals (biological, as well as non-living) to make decisions on matters that have a bearing on their daily lives. The most important kinds of information, relating to the existence of useful forest resources, are described below. Note that the following discussion highlights information that Solega people explicitly mentioned to me either during interviews or more informal discussions. Implicit signs, i.e. signs which Solega speakers do not normally articulate, are also undoubtedly very important in everyday life, and might include, for instance, the visual features of a plant that allow its identification even with a cursory glance from some distance. Solega people struggle to identify and describe such cues, just as English speakers might have some difficulty elucidating, in words alone, the difference between a 'tulip' and a 'lily'. Such matters are probably best studied by an investigator of cognitive phychology, and will not be further mentioned here (although see Excursus 6.1 for a brief explication).

The explicit signs that indicate that availability or usefulness of forest resources can be broadly divided into four categories.

a) Long-term cycles,
b) Short-term cycles and (temporal) coincidences,
c) (Spatial) collocations and
d) Intrinsic signals

[1] It should be pointed out that biosemioticians, in general, do not support ontological binaries such as nature/culture, matter/mind and verbal/non-verbal (Cobley 2010), arguing instead that culture, for instance, is merely a part of nature.

Excursus 6.1: Implicit Signs

Sitting on the rock platform called *neṭṭa kallare*, which lies a short walk from his village, M. R. Madha pointed out some trees in the distance, and casually mentioned that they were *ta:ri mara* (*Terminalia bellirica*). These trees, being deciduous, were sporting bright red leaves at the time (early May, 2009)—that was the diagnostic used by M. R. to identify this species. However, the stand of trees in question was situated several hundred metres away from the rock platform, and there was really no way in which M. R. could have distinguished *ta:ri mara* from any other red-leaved tree that might have been growing alongside them, by taking into account potentially diagnostic morphological details such as leaf shape, bark texture, and flower appearance. How, then, could M. R. be so certain that those red-leaved trees several hundred metres away were indeed *ta:ri mara*? A very important reason for his confidence was probably the fact that these trees were growing "in his backyard", so to speak. M. R. would have had ample opportunity to familiarise himself with the local tree species composition over countless trips into the forest, to the point that he could even recall the location of large individual trees that had cultural significance (for instance, if they were regularly home to large numbers of honeybee colonies). Familiarity with local species would be just one factor, however, as M. R. would probably have been able to come up with an accurate identification given the same conditions, but in a distant part of the forest. The information available to him under such circumstances would include a detailed knowledge of that particular species' ecology and phenology. M. R. would know that the *ta:ri mara* only grows in certain forest types and not others (*ka:nu oḷage e:va ja:ga:dalliwe ta:ri mara uṭṭa:dille* 'the *ta:ri mara* never grows in the evergreen forest', and also that in early spring, this tree is likely to have red leaves, but no flowers or fruit. The exact location of the tree and the time of year are therefore two examples of implicit abiotic signs (along with the more explicit "red leaves") that aided in the accurate identification of this tree species.

Many phenomena that are important to the Solega are indicated by signs from more than one category. For instance, a resource may be ready for utilisation when it undergoes a change in state (intrinsic signal)—this may usually occur in the presence of another entity (collocation) during a particular stage of the monsoons (temporal cycle).

6.2.1 Long-Term Cycles

Solega people are aware of discrete cycles of events that differ from one another in their periodicity, but otherwise demonstrate great regularity and predictability. The most extreme of these (in terms of periodicity) include the flowering of plietesials such as *kurugu* (*Strobilanthes* sp.) and *hebbidru* (*Bambusa arundinacea*). The former is said to bloom every 7 years (the most recent occurrence was in 2010), whereas the later has done so only a handful of times in living memory. Far more important than these cycles, however, is the annual monsoon cycle, which holds sway over most of the Indian subcontinent between the months of May and October. This has led the Solega to develop a calendar of rains, which cycles in close concert with the sequence of agricultural tasks.

6.2.1.1 Solega Rain Calendar

Timekeeping on an annual cycle is carried out by means of a calendar of rains, which approximates a similar Kannada system, but differs from it in significant ways. The Kannada rain calendar ultimately derives from the Sanskrit-based Hindu astrological sequence of *nakṣatra* (this term can be glossed as either 'constellations' or 'lunar mansions'), which are said to divide the plane of the ecliptic into 27 equal parts (Fig. 6.1). As there are 27 days in a sidereal month, the width of a *nakṣatra* is the arc of sky traversed by the moon in about one day. The 27 Kannada rain names closely follow the Sanskrit terms, at least where the orthography is concerned—minor deviations occur in spoken Kannada, in the lack of aspiration in the Sanskrit breathy consonants, for example. Though officially called *maḷe* 'rain' on printed Kannada calendars, these subdivisions of the year play an important role in astrological calculations. They also align with important Hindu festivals, such as (y) uga:di 'new year',[2] which coincides with the transition between *re:vati maḷe* and *ashvini maḷe*. Kannada speakers tend not to use the 27 *maḷe* names to refer to the precipitation associated with different phases of the monsoon, but instead employ two high-frequency words to divide up the rainy season—these are *munga:ru*, the oncoming south-west monsoon (June–September), and *hinga:ru*, the retreating north-east monsoon (September–November). Etymologically, *mun-* is usually associated with directional semantic content such as 'front' or 'previous', while *hin-* tends to indicate 'back' or 'following'.

The Hindu astrological almanac also has a distinct sequence of 12 months, also with Sanskrit labels, based on the sign of the zodiac that the sun transects at a particular time of year. The average Kannada speaker, however, is only aware of the English month-names of the Gregorian calendar; this is true even of people in rural Karnataka. In the Hindi-speaking parts of India, too, such knowledge is becoming more obscure, although older speakers might occasionally make reference to months such as *a:ṣa:ḍh* (mid-June to mid-July), marked by the onset of the south-west monsoons, and often associated with violent thunderstorms.

The Solega rain calendar is highly syncretic, drawing on the Sanskrit and Kannada categories described above, while also incorporating a number of concepts

[2] The Solega, while never claiming these festivals as their own, do take part in such celebrations. This is, however, a recent development, and the Solega have a completely different cycle of indigenous festivals linked to the agricultural cycle (see Fig. 6.2).

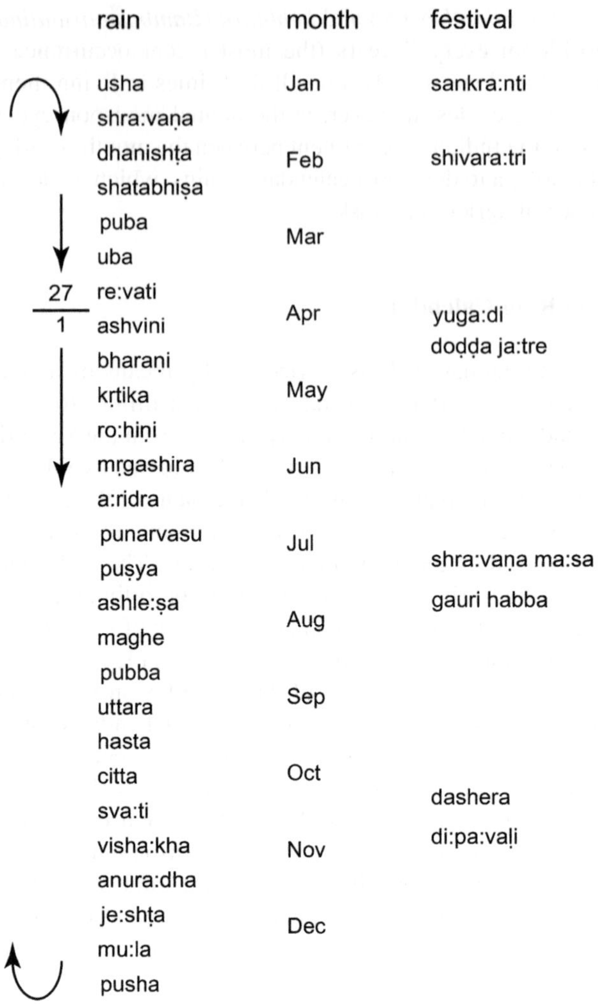

Fig. 6.1 The Kannada calendar of rains, beginning with *ashvini* and ending with *re:vati*, along with important festivals of the Kannada calendar

that appear to be indigenous. Like much of Solega traditional knowledge, the exact sequence of events in the calendar is open to a considerable amount of interpretation and discussion. One possible reason is that the calendar has long since fallen into disuse, with the result that many speakers (including older speakers) cannot now remember the full sequence of lexemes. Indeed, different people provide what seem to be different selections of lexemes from the complete list, although the ordering of lexemes tends to be consistent between speakers. People also say that the last rain in December is also the last rain of the season. This, I suspect, is not a reference to the end of the Gregorian calendar, but rather to the fact that no further rain is expected to fall during the following 3 months.

A majority of the lexemes that make up this calendar are shared with the Kannada sequence of *maḷe*, although the pronunciations are another step removed from the Sanskrit names. One significant deviation from the Kannada calendar is that the Solega sequence does not cover the entire year. Instead, the first Solega *maḷe* is said to fall in early April (Fig. 6.2), and literally refers to pre-monsoon showers, instead of a more abstract astrological notion. No rain falls in the months of January to March, and the Solega simply refer to this part of the year as *be:sage ka:la* 'dry season'. The Solega *maḷe* names therefore refer quite literally to distinct parts of the

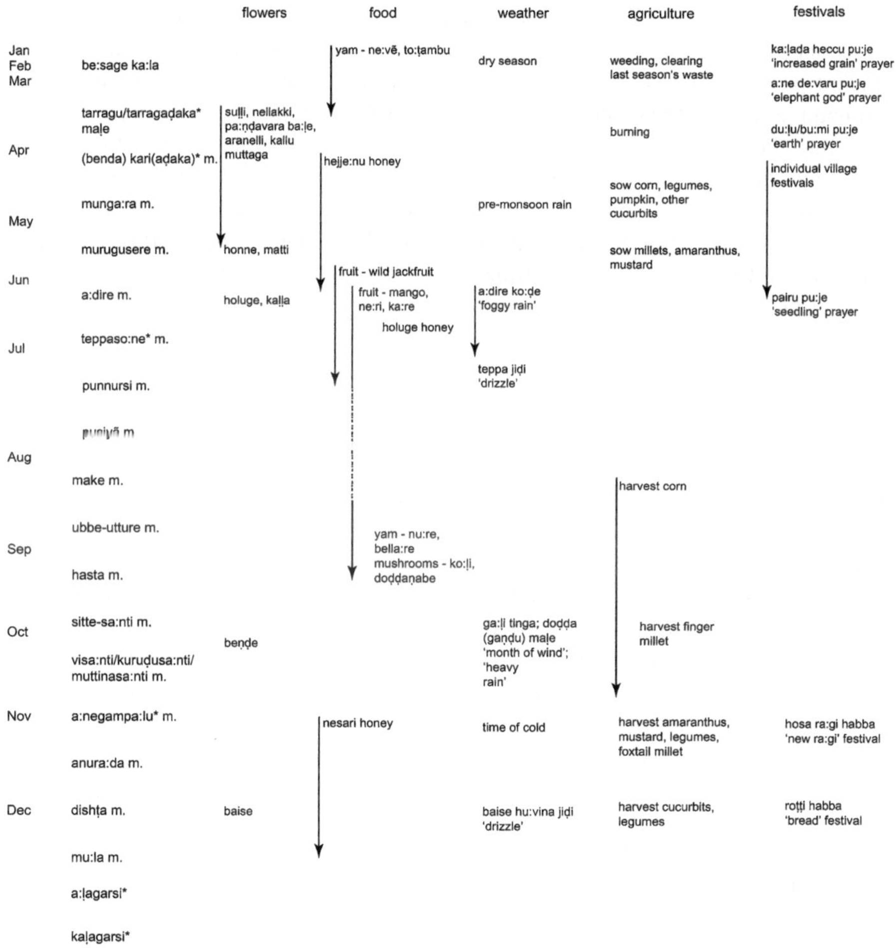

Fig. 6.2 The Solega annual rain cycle (indigenous Solega lexemes are marked with a *asterisk*), along with the significant flowers, food items, weather conditions and agricultural practices associated with each rain

monsoon cycle, and unlike the Kannada calendar, begin with the first rain of the wet season, rather than a new year signalled by an astronomical phenomenon. Another important difference from Kannada is that the term *munga:ru* has been assimilated into the Solega calendar, and here it refers to the third rain of the season, falling around early May. The Solega calendar also contains around five additional lexemes that are not present in the Kannada or Sanskrit calendars. These include *tarragu* or *tarragaḍaka maḷe* 'leaf-litter wetting rain' and *benda kari* or *kariaḍaka maḷe* '(burnt) soot-wetting rain', which are respectively the first and second rains of the season. The names encode a critical part of the Solega agriculture cycle, namely the burning of leaf litter prior to sowing.

The *a:negampa:lu maḷe* 'rain that makes elephants shiver', falling in the winter month of November, is a reference to the frequent cold and damp days that dominate this part of the year. It is entirely likely that the five 'indigenous' Solega rain terms once represented the total extent of the Solega calendar, by marking out significant parts of each annual wet–dry cycle. In this scenario, the Sanskrit-derived Kannada terms would be later arrivals into the Solega lexicon, having been incorporated into the calendar to fill the gaps between the named Solega rains. Some such rains have been grouped together in the Solega calendar, and may be considered significant because they indicate the date of an important Hindu festival. An example is *sitte-sa:nti* (some speakers also include the rain *visa:nti*) in this grouping, which is said to fall around the time of the festivals of *dashera* and *di:pa:vaḷi*. Another possible reason is that the Solega recognise this part of the wet season as a time of stormy weather (Fig. 6.2).

6.2.1.2 The Phenological Cycle

As described in Chap. 5, the blooming of important forest trees is very much an important part of Solega awareness of the forest. The spring and summer months, in particular, are times of plenty for the Solega, as all manner of fruits, which are consumed by birds, frugivorous mammals and humans alike, start to appear in great quantities on forest trees. The most important of these are *alasu* 'jackfruit', *ma:vu* 'mango' and *ne:ri* '*Syzygium sp.*', which start to become available in June–July (Fig. 6.2). The Solega also remark on the appearance of flowers on important forest trees at different times of the year—these include *honne* (*Pterocarpus marsupium*), *matti* (*Terminalia crenulata*) and *beṇḍe* (*Kydia calycina*). The times of the year when yams are available (*Dioscorea* spp.) are also eagerly anticipated by the Solega. Several wild varieties are recognised, and one variety, *to:ṭambu* 'field vine', is cultivated in home gardens alongside other crops. The cultivated yam is ready for harvest towards the middle of the dry season; most of the wild varieties are available in the late monsoon.

This phenological information neatly complements the rain calendar, and also has a bearing on the agricultural cycle. It is also used to describe the occurrence of other important biological events. Such connections between biological and climatic cycles will be discussed in Sect. 6.2.2.

Excursus 6.2: Solega Agriculture

Elements of traditional Solega agricultural practice mirror new-age, western 'organic' farming practices that are increasingly in vogue among enthusiasts of alternative, sustainable food production. The practice in question is the 'Three Sisters' method of growing corn, beans and pumpkins as companion crops in the same plot of land—this practice is often attributed, in the popular media, to the Iroquois people of North America. The benefits of this method, as stated by its modern practitioners, are threefold: that the legumes fix nitrogen for the nutrient-hungry corn plants, the corn provides the support upon which the beans grow, while the pumpkin fills the understorey, and shades the soil and roots from strong sunlight. The Solega sow *jo:ḷa* 'corn', *avare* and *togari* 'beans' simultaneously on the same cleared plot of land, while pumpkin is usually grown off to one side, adjacent to a wooden framework which acts as a support for the vine, and also keeps the developing fruit out of the reach of wild pigs. *Sa:savi* 'mustard' and *eḍḍã* 'amaranthus' can also be sown as companion crops at this time. When the corn reaches a height of *ondu aḍi* (approximately 10 inches), *ra:gi* 'finger millet' seeds are scattered between the corn and bean plants. A *guḷḷi* or temporary shelter is constructed next to the field, and the growing crops are watched over during the hours of darkness by at least one family member until harvest time. *Kaḷe* 'weeds' are removed after around 1 month, and following harvest, the grain is stored in bags within an hollow excavated in the ground. The Solega recognise different stages of growth in *ra:gi* plants, which appear as follows:

pairu	'small seedling'; common to all plants, indicates the first appearance of a seedling above the soil
ra:gi geṇṇu a:gide	'the *ra:gi* has developed joints'; indicates the appearance of distinct nodes and internodes on the developing stem
ka:sakki	a plant approximately 1.5 feet tall, starting to develop a *hoḍe* 'inflorescence'
dore	a seed head with immature grain
tene	a seed head with mature grain, ready for harvest

6.2.2 *Short-Term Cycles and (Temporal) Coincidences*

Apart from the annual cycles described above, the Solega keep track of many shorter-term biological cycles that provide them with useful information. Many of these have a period of a single day, and allow people working in the fields or in the forest to keep track of the passage of time. The calls of the *kuṭrakki* 'Barbet' and the *ka:nagoravã* 'Malabar Whistling Thrush', which sing primarily in the early morning and in the late afternoon respectively, and which make significant contributions to the soundscape of the BRT, serve to notify humans of the start and end of each working day. This is described in greater detail in Chap. 4. The opening and closing of some plants on a daily basis is also predictable enough for the Solega to use it as a time-keeping mechanism:

Ka:ḍinalli kuri oṇṭe enduru ade. Oṇṭe hu: a:gottiga, "Mu:r gaṇṭe i:ga!"
There's (a plant called) *kuri oṇṭe* (*Triumfetta rhomboidea?*) in the forest. When the *oṇṭe* flower blooms, "It's 3 pm now!" (we say).

Explicit animal signs can also be used to tell the time of day. A very familiar example is the crowing of the rooster at dawn, which in Solega is referred to as *beḷakina (in)ja:vã* 'the morning call'.

O: (su:ryaniga) ottu mu:ḍikittu, ko:ḷi ku:gottu, i: tare he:ḷta:re... ko:ḷi ku:gida:ga beḷakina ja:vã... a:ga na:ku gaṇṭe a:girutte.
"Oh! The sun has risen, the rooster has crowed," that's what they say... when the rooster crows, it's the morning call... then it's four in the morning.

Daily cycles in honeybee activity outside the hive are also meaningful, in that the lulls and peaks are said to occur at certain times of the day:

Je:nu kuruḍu edda:de, i:gatta mu:ru gaṇṭe a:gide. Je:nu kuruḍu nitta:ga sande aita:de. A:ru gaṇṭe. I:gatta kuruḍu nitta:tu koṇo, taḍe innondu ashṭottu suttumõ, a:mele ho:ga:vẽ. Innondu ga:ḷige no:ḍu alli, akka:pakka:da maragaḷa
The drones wake up, now it's 3 pm. When the drones go to sleep, it's the evening 6 pm. Now the drones have gone to sleep. Wait, let's get some more (honey), then we'll go (home). Have a look there, in the nearby trees.

The twin annual cycles of the monsoons and the flowering of plants provide an inexorably cycling backdrop before which many other biological events of shorter duration play out. These biological events achieve added significance when they are seen to regularly overlap with discrete segments of the annual cycles. Another bird, the *ke:sakki* 'Indian cuckoo', fills the air with its distinctive call towards the end of the dry season, and heralds the onset of the monsoon. As described in Chap. 7, the mass migration of *hejje:nu* 'giant honeybee; *Apis dorsata*' from the plains to the highland forests co-incides with the mass blooming of important forest trees such as *honne mara*, and is a significant event for the Solega, who then prepare themselves for the months of honey-gathering.

The naming of honey, obtained in different parts of the year, after the flowers in bloom around the time of harvesting that honey is a common feature of Solega bee-related conversation. The reason for this practice is obvious—the bees would have been foraging from particular flowers for some time before the honey was harvested, and the honey consequently takes on the distinctive aroma and flavour of those flowers. The result is a sequence of honeys that marches in step with the phenological cycle, at least for the months when honey is available for harvest—*honne je:nu* '*honne*-flower honey', *eḷakina je:nu*, *ne:ri je:nu*, *ta:ri je:nu*, *holuge je:nu*, *kaĩye kalagu je:nu* and so on. A few select honeys do stand out for being particularly delicious—*maruḷi je:nu*, which is associated with a small, unassuming leguminous herb called *maruḷi giḍa* (*Indigofera* sp.), and more generally, *nela hu:vina je:nu* 'honey from ground-cover flowers' are eagerly sought after.

There are interactions between the rain cycle and the phenological cycle, which have been lexicalised, and are used when discussing weather conditions on a finer scale. The Solega characterise some periods of rain named in their calendar as being dominated by precipitation of a certain type—for example, the term *a:dire ko:ḍe* is often used to talk about the *a:dire* rain period, as *ko:ḍe* 'rain in foggy conditions' more accurately describes the meteorological conditions prevailing at that time. Similarly, *teppa jiḍi* falls during *teppaso:ne* in July, where *jiḍi* refers to 'light drizzle', while the period of *sitte-sa:nti* in October is normally dominated by *gaṇḍu maḷe* 'male rain', which indicates heavy downpours accompanied by gusting winds. The blooming of the *baise* flower in November is taken as a reliable indicator of the approach of another type of rain—here, the precipitation is not named after the time of year during which it occurs (*i.e. a:negampa:lu/anura:da*) but after the flower. *Baise hu:vina jiḍi* '*baise* flower drizzle' is subsequently the lexicalised term for the light drizzle falling in the winter months.

Discrete points in the phenological cycle may have a bearing on agricultural terminology, once again as a result of observed co-incidences between two or more events. For instance, the penultimate growth stage of the *ra:gi* plant (when the seed-head is in the *ka:sakki* stage—see Excursus 6.2) is particularly susceptible to a disease (possibly blast), that can lead to a significant loss in yield. The *ka:sakki* stage usually co-incides with the flowering of the *beṇḍe* trees, and as a result, the disease is called *beṇḍe hu:vina visa* '*beṇḍe* flower poison' by Solega farmers. While this represents a case of a phenological event having a perceived impact on agriculture, the reverse, namely the impact of agricultural practice on phenological events, is also noted. The most prominent example in this respect is the colonisation of land cleared by fire by all manner of grasses and other pioneer species. In the days when such agricultural burning was widely practised, the resulting ash bed or *benda kari* (lit. 'burnt black') would become a rich substrate not only for cultivated crops, but also for a range of other plants that were valued for their nutritional, medicinal and aesthetic properties. These included plants like *a:sa:ḍi be:ru* (*Asparagus* sp.), said to be an effective remedy for problems of the gut; *gi:jigana ambu* (*Caesalpinia mimosoides*), whose new growth has a pleasant smell, and is used to make *sa:du*, a kind of chutney; many types of *nela hu:* 'ground-cover flowers', the source of the special honey mentioned above; and the sweet-smelling *minciga* flower (poss. *Exacum tetragonum*), a ready ally in the courting of women:

> *Ondu sari na:vu ka:ḍu benda:ga, minciga hu: andare gaṇava:da hu: buḍtittu a:ga, yelli no:ḍdare biri minciga hu:. Adu namma avaru—namma hengisaru—alli ho:da:ga avarige ishṭa bartittu, "idu ho:gu na:nu muḍiya be:ku", a:genta a: hu: muḍdu uṭṭu ka:ḍiga ho:gva:ga. I:ga hu:gaḷu ondu illa. A:ga da:rili ho:gta:idda:ga a: hu:vina ba:ri gamala bartittu…a:va:ga ka:ḍella pu:ra ondu tara gamala bartittu. I:va:ga a: gamala onduve baralla.*
> After you lit a fire, the fragrant minciga flowers would bloom, wherever you looked, there'd only be minciga flowers. Among our people—our women—they would want the flowers, [so you'd say to yourself], "I need to go pick some", and you'd do so when going to the forest. You don't get any flowers now. Back then, while walking along a path, you could really smell the perfume of the flowers… the whole forest would smell like that. It doesn't anymore.

6.2.3 (Spatial) Collocations

The knowledge of 'what is found where' forms the core of any human's understanding of his/her environment. One would expect people living away from urban centres to be well aware of the preferred habitats of plants and animals, and the Solega are no exception. Chapter 5 describes in some detail the Solega's knowledge of the plant assemblages associated with different topographies, climatic conditions and substrates. In this context, elephants are said to prefer certain types of vegetation during particular seasons, and this has an important impact on the free movement of humans through the landscape. Their knowledge of bird and smaller animal species distributions is equally fine-grained, and takes into account potentially complicating behaviours such as feeding, reproduction and migration patterns. The following example deals with the habits of the *ko:kunji* 'poss. Slaty-legged Crake':

> *Ko:kunji bu:miliye. Ṭīyanakki tara, kallu se:rsukoṇḍu alli mari kaṭṭade. Haḷḷada kaḍegaḷalli ka:ṇta:irtade.*
> The *ko:kunji* stays only on the ground. Like the lapwing, it collects stones, and has its young there. You can see it around the banks of streams.

The Solega recognise several prominent *je:nu mara* 'bee trees', which can be home to 40–50 individual giant honeybee colonies (see Chap. 7 for further discussion). This represents one of the most prominent examples of a reliable spatial collocation between two organisms. Another spatial collocation that used to be far more common in the days of burning (and consequently, the days before the *Lantana camara* invasion) is the presence of *nela je:nu* 'ground-dwelling bees' (actually instances of *Apis cerana* nesting in abandoned animal burrows) close to patches of *nela hu:* 'ground-cover flowers', the source of the special honey of the same name. Both organisms are now rare due to the massive proliferation of *Lantana* on the forest floor, which has caused bees to mainly nest on/in large trees.

Mushrooms are plentiful in the BRT forest, but the Solega only name and eat a small subset of these organisms. Here, too, spatial collocations are highly meaningful, and operate at two distinct levels. At the macro-level, it is important to know that the *doḍḍaṇabe* 'big mushroom' and *eṇṇe aṇabe* 'oil mushroom' are found in dry clearings, grassland, cultivated areas or burnt patches of ground, whereas species such as *ko:ḷi aṇabe* 'chicken mushroom' are found in moist, densely forested areas. These three mushroom types grow out of the ground, whereas *beṇḍe aṇabe* and *karava:di aṇabe* sprout from rotting tree logs. Knowledge of spatial collocation at the micro-level is necessary at this point, as the latter two mushrooms are only said to grow on the rotting logs of the *beṇḍe* (*Kydia calycina*) and *karava:di* (*Persea macrantha*) trees respectively, after which the mushrooms are named. The Solega are wary, for good reason, of eating mushrooms they do not recognise and/or name. The two wood-mushroom species may well grow on other substrates, while other mushroom species very likely grow on the logs of the two trees named above. However, the prescription of a tight, formal association between a mushroom of a particular appearance and a particular substrate appears to be a reliable strategy to ensure that only culturally-sanctioned mushrooms will be consumed.

> **(K)** *Kaḍinalli ja:sti beṇḍe maragaḷu ide. Adu a:negaḷu u:ḷakke murda:kurtave. Murda:kida:ga adakke ondu eraḍu varsa ṭaim a:gira be:ku. Oṇagira be:ku. A: ṭaimalli maḷe biḍutte, ju:n julaili. Ava:ga maradindane aṇabe baratte... Karava:di aṇabe ade tara... Be:re maradalli aṇabe barutte, adu na:vu tegeyalla, ya:kandare adu be:re mara.*
> There are many *beṇḍe* trees in the forest. Elephants push them over while foraging. After they get uprooted, they need to lie there for a year or two. They need to be dry. After that the rain falls, in June or July. Then, mushrooms appear out of the wood itself… *Karava:di* mushrooms are exactly the same… You see mushrooms growing on other trees, but we don't take those, because they're different trees.

In the same way, one presumably feels more at ease collecting *doḍḍaṇabe* from around a termite nest, because that is where they are said to usually occur:

> **(K)** *Eṇṭu-ombatne tingaḷalli solpa ka:ḍu ja:sti ira ba:rdu... A:mele solpa o:pan ira be:ku. A:mele ondu tara uttagaḷu barutte alli. Uttagaḷa pakka, adara sadurane... doḍḍaṇabe barutte... Benki a:da:ga, ka:ḍu benda:ga sigutte. Jami:nalli uve solpa sigutte... Oḷage gejalu irutte, gejalinda sekkana:gi bara be:ku, ashṭe. Maṇṇindave srushṭi a:ga be:ka:du.*
> In the eighth or ninth month, (in a place where) the forest isn't very dense. It should be a bit open. That's where you find termite mounds. Near the mounds, all around them… you find *doḍḍaṇabe*. When there's a fire, after the forest burn, you find them. You also get some from a field… There are termites inside (the ground), and (the mushrooms) grow well around termites. They arise straight from the soil.

6.2.4 Intrinsic Signs

All organisms, including humans, give off signals that can be received and interpreted by the sensory organs of other organisms. Such signals may be emitted actively or passively, and can convey useful information pertaining to the presence or the current state of the emitter, or to a change in state, among other things. Solega people make liberal use of such signs, and are able to talk about them explicitly, in sharp contrast to the implicit signs that determine the identification of, and discrimination between, different organisms. For instance, the *doḍḍaṇabe* 'big mushroom' mentioned earlier are only good to eat at certain points in its life cycle, and these are indicated by its physical appearance:

> **K:** *Na:vu modalu no:ḍda:ga [doḍḍaṇabe] ondu mogaṭu tara baratte. Mogaṭu tara banda:ga na:vu tagoḷtivi... mogaṭu tara bandu adu fullu ra:uṇḍ a:gutte, adu na:vu agalide endu he:ḷtivi. A: tara banda:ga adara oḷa:ge ondu huḷa bandu uḍte, a:ga na:vu tagoḷalla.*
> When we first look, [the big mushrooms] are shaped like a crown. When they're that shape, we pick them. After they become crown shaped, they turn fully round, we then say that they have opened. When they become like that, they're full of bugs, we don't pick them then.

Possibly the most important explicit animal signs for the average Solega are those that indicate the presence of elephants nearby. Solega speakers often go into various forest habitats frequented by elephants in order to forage, collect raw materials, or tend their fields. Elephants will also wander close to human settlements to raid grain fields and banana plantations, and to drink from artificial reservoirs in the dry season. An unexpected encounter with an elephant can very easily prove fatal, and the Solega are always on the alert for the primarily auditory cues that signal the presence of elephants:

> *Ondu maravo murritave, "bu:r" endave, ada hiḍiya bahudu. Matte kĩvĩya ondu taradalli "moṭṭakko moṭṭakko moṭṭakko" endu hoḍda:de. A: saddadalli hiḍitivĩ na:vu. Ondu ondu a:ne alli murda:de, adave murda:de. "Kirri" enda:de, idalli "goḍrrr" enda:de, e:vadondu sadda koṭṭa:de. I: he:ḷu "boddo boddo boddo boddo" endu surda:de. A: tara nanaga gotta:gi "idu a:netta, be:re oṭṭo:gõ", enda:ki oṭṭo:itivĩ.*
> They'll shake a tree, you hear "*bu:r*", that's how you know. Then they flap their ears, making the sound "*moṭṭakko... moṭṭakko... moṭṭakko*". That's how we know they're there. The elephants trumpet, going "*kirri*" or "*goḍrrr*", one of those sounds. They make the sound "*boddo boddo boddo boddo*". That's how we know, "There are elephants here, lets go elsewhere," and we go away.

While the sounds made by elephants tell a human that the animals are dangerously close, visual signals serve to provide an earlier warning of an elephant's whereabouts:

> *Hosa ajje biṭṭide, o: illi a:ne ade, illi ho:gu be:ḍi i: gaddega endivi. va:saneli gotta:itade, jiḍḍoḍetade, laddi ha:kirtade, i: mara giḍa ella murdu tindukoṇḍu ho:girtave, inta ja:gadalli a:ne ade*
> There are fresh tracks. "Oh! There's an elephant here, don't go into this forest!" we say. We know from the smell—the elephant urinates, defecates, it breaks branches and tramples plants as it passes along, eating. Places like these have elephants.

Sometimes, however, one finds onself face-to-face with an elephant. In such a situation, it is important to try and predict what the animal might do next, by looking for crucial signs:

> *Ti:rtade ka:la, munga:la ti:rtavane avã, avã ro:puga:rã endu. A:ga na:vu tappisuma:ku.*
> It scrapes the ground with its foreleg, with its foot, to let you know that it's angry. That's when you need to flee.

Humans also give off signals that may be picked up by other animals. This category of signs was rarely mentioned in my recordings, but two instances of animals reacting to the presence of humans were mentioned by a few speakers. In both cases, elephants were the recipients of both physical and 'metaphysical' signs emanating from humans, and reacted appropriately. In the following extract, the speaker explains how elephants are able to detect the presence of humans in the forest, even before the human might be aware of the elephant's proximity:

> *Namma va:sane a:neka gotta:dde. A:neka to:rrda:de, a: va:sane. A: va:sane to:rrda:ga a:ne alli suṇḍilu etti a:ga no:ḍda:de. Adakka gotta:go:da:de, namage gotta:ga:dille.*
> The elephant knows our smell. It alerts them, and when it smells us, it lifts up its trunk, and looks around. That's how it knows; you wouldn't know that it's there.

Elephants also react to the taint of wrongdoing emanating from one who has broken social or religious norms, and can unleash the ultimate divine retribution upon the offender:

Excursus 6.3: Elephant Names

Solega has around 25–30 lexemes for elephant types, based on age, gender, tusk morphology, or combinations of two or more features. The following image shows a handful of these elephant terms—such terms are frequently used in naturalistic speech, as it is customary to indicate, when describing encounters with elephants (or informing others of the presence of elephants nearby), the type of animal being talked about.

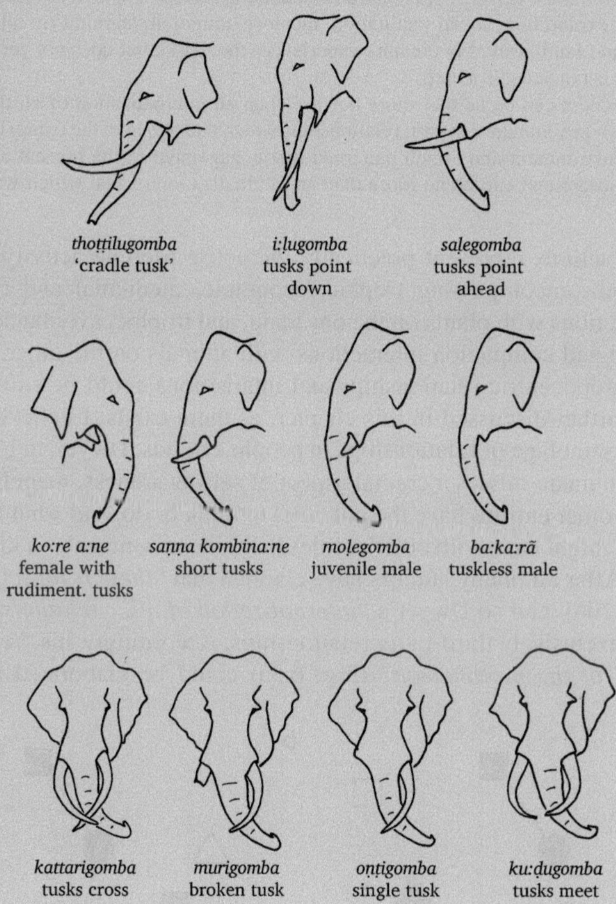

Solapa inkura tappu sikkittu enda:ga a:ne elli iddarue bandu uṭṭadde, avana se:rika. A:ga bandattu endale gedda:du gela:dondu, avã sattale ondu, tappu bandale ma:tra. Tappu ille endu o:tu enda:ga alli biddareluve a:ne e:nu ma:ḍa:dille.
When an elephant detects even a bit of wrongdoing, it will turn up to meet [the offender], no matter where he is. Once it arrives, it may kill him, but only if he has done wrong. If he is innocent, the elephant will not do anything, even if he falls (while trying to run away).

6.3 Relationships

The reception and interpretation of meaningful signs is complemented by the recognition of real and potential relationships between oneself and another entity in one's environment, or between two non-self entities in one's environment. The former type of relationship has been presented as a pre-requisite for the quality of 'self-awareness' that our species acquired while developing from 'pre-human' to 'proto-human' [14].

> Imagine a pre-human being… A creature embedded in relationships, relationships that are grounded in exploration, social facilitation, memory, imitation, learning (in all its forms), instruction and familiarity. My creature experiences the world, but does not yet imagine it. My creature is not yet self-aware.
> … self-awareness can be neither more nor less than an internalization of all the relationships of which pre-human was part; relationships which had arisen in the context of engagement with environment and which patterned that engagement in the present and into the future. Self-awareness can be no more than an internalization of that which was external. (pp. 18–19)

Such interactions represent practically the entire lifetime activity budget of a Solega person—encompassing trophic, economic, medicinal and construction-related interactions with plants on the one hand, and trophic, avoidance, antagonistic, economic and companion interactions with animals on the other. Many other types of anthropocentric relationships and interactions could be enumerated, but will not be further discussed in this chapter, as there exists, I believe, a far more interesting assemblage of relationships in people's minds. Dwyer, in his discussion on the proto-human, misses a crucial aspect of self-awareness, namely the knowledge of what other entities have the potential to think or do, and what relationships those entities might have with other entities in their environment—a kind of 'social awareness'. After all, many authors have claimed that "*there is no self without the other*" [205, 206], and so Dwyer's "*internalization of all… relationships*" should also include exclusively third-party relationships. Accordingly, his "*representation of the world of the protohuman*" (Fig. 6.3a) could be elaborated as shown in

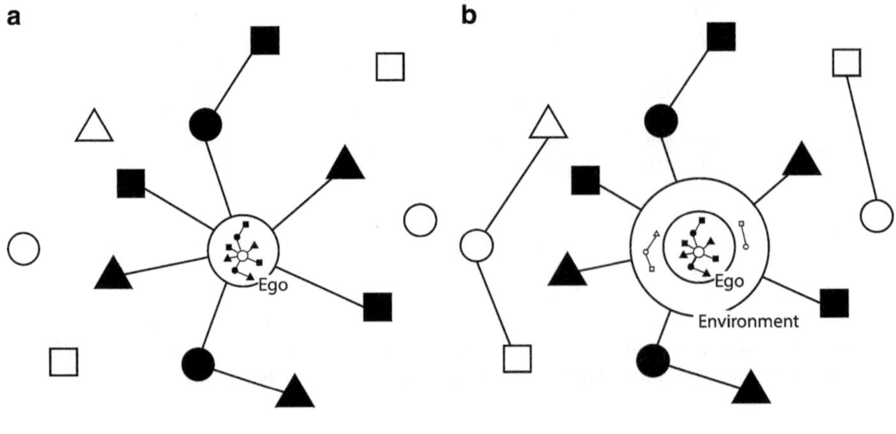

Fig. 6.3 (**a**) Representation of the world of the protohuman, according to Dwyer (2005). (**b**) Dwyer's representation modified to include knowledge of third-party relationships

Fig. 6.3b. Here, 'self awareness' comprises knowledge of not only the ego's relationships with other entities, but also of relationships in the ego's environment that do not immediately involve the ego. Of course, recent advances in the fields of ethology and behavioural ecology have demonstrated that these phenomena are not exclusive to humans, and that there are plenty of other (for the moment, vertebrate) species that are aware of both types of relationship in their environment (e.g. [207, 208]). Nevertheless, it is still pertinent to ask why humans would want to attend to relationships that do not concern them directly. It has already been suggested in Chap. 3 that the knowledge of plant–animal interactions allows people to find animal prey, as well as avoid dangerous beasts (also see Sect. 6.3.1 below). Here, I suggest a further hypothesis to explain the phenomenon of attention to third-party relationships after describing some such relationships that Solega people are aware of.

The vast majority of third-party relationships (at least for plant–animal and some animal–animal pairs) that Solega people speak of are trophic, or have a trophic component, although non-trophic antagonistic relationships between animal or plant pairs are also frequently mentioned. Each of these pairs will be discussed in turn.

6.3.1 Plant–Animal

'Animals (and birds) eat plants' is the central theme of this pairing, although there are many variations thereof. The feeding habits of elephants figure prominently here, due to their massive size, and their ability to cause very visible damage even to large trees in their search for food. Elephants are known to knock trees over or grab hold of tree trunks and shake them, in order to get at ripe fruits. They will also pull bark off living trees (commonly *daḍasina mara*, *matti mara* and *sirhonne mara*) with their trunk or feet and consume it, and frequently scratch their backs or hindquarters on tree trunks, the underside of low branches or large boulders to relieve an itch. Clear telltale signs of the latter activity, in the form of smoothened, mud-smeared tree trunks and branches, can be easily spotted in the forest. Other herbivores such as various deer species will also scrape bark off trees with their antlers.

As mentioned in Sect. 6.1, when asked to identify a plant, Solega people tend to also spontaneously mention the birds and animals that use that plant as a food source. Ferns are quite aberrant in this respect, as no herbivore is known to feed on their leaves. The vast majority of named plants, however, form a link in at least one food chain, which may or may not involve humans. In any case, there is considerable overlap, as the fruits and foliage of many plants are food for humans and animals alike.

Table 6.1 gives a list of some of the preferred plant foods of prominent forest animals and birds. The consumption of fermenting *doḷḷi* (*Careya arborea*) fruit by elephants in mid-summer (June) was frequently mentioned to me, as it invariably has the comic, and sometimes dangerous effect of causing drunken behaviour

Table 6.1 The preferred plant foods of various animals

Animal	Plant food
a:ne 'elephant'	*ba:ṇe, na:ralu, ko:ḷi* grasses, fruit of *doḍḍa ka:re, doḷḷi* trees, fibres of *beṇḍe, daḍasu* trees, *kaũri* vine
ka:ṭi 'gaur'	*ba:ṇe, na:ralu, joṇḍu, sabbe, ko:ḷi* grasses
kaḍave 'sambar'	Various leaves and grasses, *ku:ḷi* flowers and fruits, *ne:ri* fruit, unripe *si:ge, aṇṭuva:ḷa* fruit
kuri 'barking deer'	*ne:ri* fruit
ku:randi 'chevrotain'	Fruit of *ku:ḷi, ne:ri* trees
karaḍi 'sloth bear'	Fruit of *ciriporika* plant and *gokkoruḷe* vine
ka:ḍabekku 'civet'	Various fruit
o:ṭe handi 'pangolin'	Flowers of *a:sa:ḍi* plant
handi 'wild pig'	Bulb of *ku:re pandi* plant
a:regã 'flying squirrel'	Unripe *matti, bejja* fruit, young *doḷḷi* leaves
kuṭrakki 'barbet'	Fruit of *kara:va:di* tree
Birds	Fruit of *ta:re, icci, ne:ri* trees, *ra:gi* grain

in the animals. Even more interesting are Solega observations on what appear to be self-medication behaviours in elephants for a variety of ailments, ranging from dizziness to pathologies of the alimentary tract to peri-natal complications. Such complaints are infrequent, however, as the Solega say, "*A:neka ro:ga baradu kammi*" 'Elephants seldom get sick'.

> *A:ne basuri a:gide…herige daṇḍe aitade enda:ga sirhonne sekke tandu, adava karku-karku endu tindu, nungu uḍtu. Nungida:ga herigeli ye:nu tondari a:ga:dille. Sali-sali herige aitade. Innu, karaḷunalli e:na a:girtade… e:vadondu no:vu bande bartade. Banda:ga hoṭṭe oḷage e:vadondu gaṇṭu irtade. Idda:ga adakka gotta:itade a:ga. "E:vadondu nanaga ka:yalu ade" endu ho:gi ondu beṇḍe marada tokkeya iḍidu, a: tokkeya tinda:ga—beṇḍe tokke biri lo:ḷi—adava a:ne tindade, matte kaḍave, kuri elegaḷa tindave… a: ele tinda:ga i: be:re ja:ti pera:ṇigaḷa hoṭṭeli ira ka:yalu kaḷadodde… Tinda:ga lo:ḷili kaḷadodde.*
> An elephant falls pregnant… when the time of birth draws near, it takes some *sirhonne* bark, munches it up, and swallows it. Then there are no problems with childbirth. Childbirth proceeds smoothly. Next, something may happen to its guts… it may feel some discomfort. When that happens, there may be a blockage in the gut. At that time, it knows, "There's something wrong with me," and it goes and grabs some *beṇḍe* bark, and eats it—*beṇḍe* bark is very soapy—elephants eat it, and the sambar and barking deer eat its leaves… when they eat it, whatever's wrong with the guts of these animals goes away. The soapiness takes it away.
>
> *Innu be:di a:dde a:neka. A:nega be:di a:da:ga, aranelli marada sekke tinda:de. Matte kaũri giḍuvina tokke bari togaru. A: kaũri na:ra murtu tindu uṭṭale a:neka be:di nindottu. Adara uccalu nindottu. Hi:ge a:neya aũsti adu. Innuve adaka be:di sariya:gi nilalle endale, innu ondu kelsa ma:ḍda:de. I: uppu maṇṇu, lakke maṇṇu endu ade… ond-ondu ja:gada maṇṇu, rusi a:girtade. Tindare gamm-ennadu. Ondu tara ba:eke rusi to:radu. A: maṇṇa ho:gi ondu eraḍu ba:e ka:linalli oddu, adava tindu uṭṭale a: be:di a:ga:du nintodde… Huḷu bandu uḍtade, a:ne hoṭṭeli. Adakka i:ga e:va sariya:da aũsti sikka:dille. Bidiru soppu be:ku, kiribidiru. Adu i:ga sikka:dille. Kiribidiru soppu, adu oḷḷedu a:neka.*
>
> Next, elephants get diarrhoea. When that happens, it eats the bark of the *aranelli* tree. And the bark of the *kaũri* plant is very pungent. When the elephant breaks off some *kaũri* fibres and eats it, the diarrhoea stops. That's elephant medicine. If the diarrhoea still doesn't go away, it does another thing. That salty clay, it's called *lakke maṇṇu*… you find it in some

places, it's tasty. If you eat it, it's yummy! It brings quite a taste to your mouth. It goes and digs up one or two mouthfuls of that clay with its foot, and when the elephant eats it, its diarrhoea stops… It can get worms in its belly. There's no good medicine for that nowadays. You need bamboo leaves, leaves of the dwarf bamboo. You don't find that anymore. Dwarf bamboo leaves, they're good for elephants.

Makkaḷu a:ga:dakka, sette bardille. Adu taḍdu irtade, hoṭṭe oḷage. A:neka ue a:gira bahudu, be:re pera:ṇika ue a:gira bahudu. Ag:a si:re soppu endu ade, si:re endale bidiru soppu, kiribidiru soppu. A: kiribidiru soppa ho:gi tinda:ga horagaḍe bartade. Innu goṭṭi sekke ade—a:neka ue a:gali, namaga ue a:gali, kaṇṇige ka:vaḷa, tale suttu bartade. Tale suttu bandale goṭṭi sekke tinda:ga suttu nindottu. Matte a:ne e:vadondu ti:niya tinduru ho:ga ho:ga, a:neka ondu tara sokku irtade. Huṭṭida mansaniga ka:yalu bardiddada? Bande bandade. Ha:geve ha:gondu ka:yalu a:neka banda:ga adu si:ge paṭṭe eḍdu—sokku si:ge—adara amba tindu uṭṭale a:ga adakka bandira manku oḍdottu.
At childbirth, the placenta is [sometimes] not expelled. It remains stuck within the belly. It can happen to elephants, it can also happen to other animals. There's something called *si:re soppu*, it's just the leaves of the dwarf bamboo. If the animal goes and eats some, the placenta is expelled. There's also the bark of the *goṭṭi* plant—sometimes your eyes go dark, and your head spins. It happens to us, and it happens to elephants. When elephants feel dizzy, they go and eat some *goṭṭi* bark, and the spinning stops. And sometimes elephants feel weak, as they walk along. Don't humans fall ill in the same way? They certainly do. When an elephant gets something like that, it peels off some bark of the *si:ge* plant—the *sokku si:ge* plant—if it eats that vine, the drowsiness that has descended upon it goes away.

The relationship between animals and plants can sometimes be antagonistic, and Solega children are warned by their elders not to touch plants like *guḍsã* and *cikka ku:guri ambu*, whose seeds may cause one's skin to itch or blister. Elephants can also be affected by plants such as *Lantana*, whose thorn-lined woody stems create barriers so impenetrable that even these large mammals are reluctant to walk through. Sometimes, an elephant simply has no choice, and the results are unpleasant:

Ku:guri ambu endale ba:ri oḷḷe u:ṭa adakka. Adu la:nṭa:na me:le ho:girutte. La:ṇṭa:na me:le ho:da:ga tinnakka:galla. Adakka pu:ra la:nṭa:na oḷa:ge nuggutte adu. La:ṇṭa:na oḷa:ge a:ne ho:da:ga muḷḷella pu:ra eḷta:ku uḍte, mai ella. Adara mai ella pu:ra ga:ya ma:ḍu uḍte, a:neka. A:va:ga idda a:neka uve i:va:ga idda a:neka uve vetya:sa irutte. E:kandale a:va:ga o:panalli o:ḍa:ḍtittu, i:ga tondari pu:ra adakka. E:kandale i:ga tinna be:ku enda:ga illinda ti:nika kashṭa:gi ho:ga be:ku.
The *ku:guri* vine is something they really like to eat. It grows over the *Lantana*. When it does that (the elephant) can't eat it. That's why it must go right into the *Lantana* bushes. When an elephant goes into *Lantana* bushes, its skin gets scratched all over by the thorns. The elephant gets cut up quite badly by them. There's a difference between the elephants of the past and the elephants you see today. Because back then, they'd run around freely in the open, but now they have a lot of problems. Because now when they want to eat, it's really hard for them to get to the food.

Solega knowledge of the trophic habits of elephants is rivalled only by their knowledge of honeybees. As mentioned above (and in Chap. 7), the temporal relationships between migratory honeybees and certain flowers are known to the Solega, as are the links between variously-flavoured honeys and the mass flowering of particular plants. Finally, the significance of known (and named) *je:nu mara* 'bee trees' to the Solega cannot be overstated, as these represent an important, annual source of both income and calories.

6.3.2 Plant–Plant

The Solega are aware of the differing nature of epiphytic and parasitic plants, and of the effect they have on the host plant. Orchids are a diverse, common and conspicuous group of epiphytes in the BRT, but as they have no noticeable impact on the health of the host trees, and as they are not utilised for any purpose by the Solega, they are largely ignored, and the individual species are not even named (see Chap. 2 for a more detailed discussion). Their epiphytic habit is referenced in the name of the group as a whole: *mara banja* 'tree *banja*'. In contrast, the effect of *uppilu* 'mistletoe' (*Taxillus tomentosus*) on its hosts is seen as a negative one, especially if that host is the culturally and economically important *nelli mara* 'Indian gooseberry or *a:mla*' (*Phyllanthus emblica* and *P. indofischeri*). The current bans on burning have, in the eyes of the Solega, increased the parasite load of *uppilu* on individual *nelli* trees, with the result that yields of fruit are now far lower than they were in times when agricultural burning was carried out. The results of recent scientific studies have suggested that this may well be the case [209].

Lantana leaves its mark not only on the hides of elephants, but also on the population structure of the forest. Solega people have noticed that once common medicinal and food plants, along with mushrooms and numerous fragrant, ornamental flowers are now hard to find, because the *Lantana* bushes simply leave no space for potential competitors to grow. There are also far more menacing, longer-term consequences of the invader's presence—the worst affected patches of the forest have now lost their regenerative capacity, and face imminent destruction:

> *I:ge:n a:gide? Marada taragu be:ya ida ottiga me:le ishṭu ga:tura bitta biddaruve taragina me:leve, ondu bitta uṭṭa:geuve da:ri ille. A:ga maḷe biḍta biḍta ni:ru biddu koḍte ottu, i:ga giḍa bardille. Saṇṇa giḍa ille, matte saṇṇa pairu illa matte e:n andare—ro:ja ma:tra. Jo:ra:goitu; arda mara gaṭṭa uve! … Marada geḍḍeka ni:re ille i:ga, be:re hotteve tindurtu adava. Mara ella oṇa:gitta biddiade. I: ka:lakka na he:ḷa:du, mara ella pu:ra oṇa:gi ottu, saṇṇa mara illa doḍḍa doḍḍa mara:de. I:vottu benki biddottu enda:ga, mara ella ta:ma:eya ho:tu i:ga.*
> What's it like now? When the leaf litter isn't burnt, all the seeds fall on the litter, and not a single one is able to sprout. And then the rain keeps falling on the seeds, and they all just rot. No small plants get to grow—just the *Lantana*. There's lots of *Lantana*; the bushes grow as tall as half a tree's height! … So now the trees' roots don't get any water, because the *Lantana* bushes drink it all. All the trees dry up, and fall over. I'm telling you, at the present moment, all the trees are drying up, and no new trees can grow, so the forest's full of big, old trees. And so when there's a forest fire, all the trees are destroyed.

6.3.3 Animal–Animal

Relationships between animals include trophic interactions, as well as agonistic and antagonistic exchanges. Obvious examples of the former include the carnivores of the region, including *huli* 'tiger', *kirubã* 'leopard' and *kenna:yi* 'dhole'. The two larger solitary carnivores are rarely encountered by the Solega, although they frequently attribute the disappearance of goats and domestic dogs to attacks by these predators. Some birds are also implicated in interactions with other animals, including predation in the case of the *ko:ḷi koṭã* 'chicken hawk' (Sparrowhawk), mild, unintended antagonism towards elephants in the myth of the *kuruḷihakki* 'quail' and governance in the case of the *doḍḍa karaḷi* 'Racket-Tailed Drongo'. All these cases are described in more detail in Chap. 4.

6.4 'They're Like Us'

The majority of third-party relationships that Solega people notice as being present in their environment are plant–animal (mostly trophic) interactions. But why should any group of people pay so much attention to a host of unrelated phenomena that—it could be argued—have little or no impact on their own lives? Why should, say, an elephant's eating habits be the type of information that is attended to by humans going about their own activities, information that is often the topic of discussion, and that can be effortlessly recalled at will? Shouldn't egocentric relationships be of paramount importance in a human's self-awareness, as suggested by Dwyer's model (Fig. 6.1a)? Reiterating two competing arguments that I discuss in Chap. 2, I could ask whether it is the elephant's physical size that causes people to notice its every move (with either detached interest, or with caution)—i.e., perceptual reasons—or whether it is the possibility that humans might learn valuable lessons from such observations, regarding the edibility or medicinal uses of certain forest products—i.e., utilitarian reasons. From my discussions with various Solega people, I suspect that although both the above arguments might provide partial explanations, there is a third possibility that provides the most satisfying answer. Simply put, Solega people notice third-party interactions in their environment, because they empathise with the birds and animals that form a part of those interactions, noticing several lines of similarity between their own lives and those of other organisms. The 'elephant medicine' transcripts presented earlier show that the speaker unambiguously equates the animal's maladies with familiar human ones, although the remedies for humans involve the ingestion of different plant products. On the topic of health and well-being, the same speaker goes on to deliver the following message on the importance of salubrious living, once again drawing strong parallels between the human condition and the life of animals:

> *E:va miruga baduka:giddaruve, a:ne baduka:giddaruve, na:vu namma maneli e:va ri:ti kli:na:gi haccu kaṭṭa iddamõ, a: ri:ti adakka mane haccu kaṭṭa ira:ku. Na:vu e:ka ke:ri guḍsadu? I:ga ni: bande, naṇṭa banda:ga ni: ke:ḷde, "E:na? Ivara ke:ri hi:ge ade?" endu heḷtive! Na:sadillava? A:ga ma:naka kammi. Ha:geve, a:neka uve, mirugaka uve, matte prationdu ma:nava jaluma, iruppu embattu ko:ṭi, e:na ji:va ra:shi huṭṭiddado, a: ji:va ra:shika ella sampu:rṇava:gi ira be:ku. Sampu:rṇava:gi iddareve namage ro:gagaḷu baradille.*
> When an elephant or some other animal is fully grown, it likes to keep its home in order, in the same way that we like to keep our house tidy. Why do we sweep the floor and clean our house? Someone like you comes along—a guest—and you think to yourself, "What's this?

His house is such a mess!" Wouldn't that be terrible? How shameful! In the same way, elephants, other animals, all humans that have ever lived, be it the 90 million ants that crawl on the ground or any other form of life, all living beings need to keep their life in order. If everything is in order, diseases will not affect us.

Solega speakers do not hesitate to make use of anthropocentric analogies to help explain the behaviour of all manner of animals. One might suspect that only mammals are so privileged due to their overall similarity to humans, but the following extract shows that human-like qualities may even be attributed to invertebrates, such as the honeybee:

Ade meneka adu gurtade. Adu elli hoṇṭo:girtade, puna banda:de. I:ga, ni:vu illi ivire. Illi iddu ni:vu oitivire. Ho:gu uṭṭu bara be:ka:dare elli bara be:ku? Alligave bara be:ku. Ha:geve je:nina kombu—a: je:nu bandu mudi tu:kidadella, a: tara ni:vu. A: je:nu illi ku:turu, a: tuppa ella se:rsi, idella u:ṭa ma:ḍu uṭṭu, tuppa ti:rtade enda:ga oṇṭodde. Oṇṭo:gi matte puna e:va ṭaiminalli bara be:ku, a: ṭaiminalli bandu ku:tumadu. Ashṭu kelsa je:nu ma:ḍa kelsa—mansa he:ge ma:ḍdade, aduve ade tara ma:ḍdade. I:ga na:vu ku:turu ma:ta:ḍtivĩ, mu:r jana uve. Na:vu edduru oṭṭoitivĩ, puna innondu jina banda:ga, elli bandu ku:tõ? Alligave—je:nu kaṭṭa kombigave!
It remembers its own home. It goes away somewhere, then comes back. Now, you're right here. After a while, you go away. If you want to come back, what do you do? You need to come right back here. It's the same with a bee's [home] branch—the colony comes and creates a new home, just like you. The bees sit there, collect honey, eat their food, and once the honey's gone, they fly away. And when they need to come back some other time, they return and land [there]. That's what bees do—they do exactly what people do. Now, the three of us are sitting here, having a chat. We get up and go away, but if we want to meet another day, where do we go and sit? Right here—back at the bees' home branch!

There are other lines of evidence that support the hypothesis that Solega people really do identify with a range of non-human organisms—examples can be found, in Solega mythology, of organisms that are said to possess human-like traits. In some cases, this is the result of an organism once having been a human, who was forced to take on a non-human form due to extenuating circumstances. Trees were once humans who had avoided offering any aid to the god *ma:desurã* during his battle with the demon *savaṇa*. Seeing that these same people expected to profit from *savaṇa*'s death in spite of their inaction, *ma:desurã*, in a fit of rage, cursed his cowardly followers, and turned them into trees where they stood. In another story, a woman was charged with caring for her aged in-laws while her husband was away, but was woefully negligent in carrying out her duty. Upon his return, the husband found his parents in a poor state of health, and cursed his wife, turning her into a dog that fled before his wrath. A third story tells of a woman who was treated poorly by her mother-in-law, being asked to perform a laborious task that the latter had ensured could never be completed. The woman was left working at home, while the mother-in-law put on her finery, and set out for a festival. The woman, seething at being tricked, used some household objects, and turned herself into an elephant—a winnowing fan for ears, a long pestle for a trunk, four grinding stones for feet, a bamboo barrel for her body and a broom for a tail. She hid in the forest by the path that her mother-in-law was bound to take, and upon seeing her returning from the festival, threw her to the ground with her trunk, and trampled her to death.

Incidentally, this last story also explains why elephants are frequently ill-tempered when encountered by humans.

Solega farmers are stoic about the reduction in grain yields brought about by crop pests such as parrots and wild pigs. They do take measures to keep animals from their fields—by building shelters beside their land, making fences with thorn bushes, and keeping watch at night, for instance, and they do suffer keenly the loss of a year's labour at the hands of a band of elephants. However, there is a tendency to perceive such events as natural, albeit unwanted, phenomena—events precipitated by the all too familiar human condition of being hungry.

> **K:** *Ondu e:kareli handigaḷella, be:re pra:ṇi ella tindu namma guḷi oḷage heccu kaḍime ondu eraḍu mu:ṭe ra:gi baratte. Arda pra:ṇige ho:gutte, arda namage sigutte.*
> After the pigs and other animals have eaten, one acre yields about one or two sacks of grain to put into our storage pit. Half goes to the animals, and we get the other half.

The farmers understand that critical changes in the forest ecosystem are to blame for the actions of the elephants:

> *Ippatu varshadinda a:kaḍe, ka:ḍu ja:sti o:pan ittu. A:va:ga i: a:ne, ka:ṭi, kuri, adakke ella oḷḷe u:ṭa siktittu. Adarindave ya:va kaḍe ue janagaḷige tondari koḍtarililla, athava holagaḷige bartarililla. I:va:ga la:ṇṭa:na ja:sti a:goitu. Adakke alli u:ṭa sigalla. Adarinda jami:nu kaḍe o:pan iruttella? A: ja:ga senda:gade, adara oḷage bandu uḍte. Banda:ga—na:vu illi alasina mara ha:kirtivi, a:mele ba:ḷe ha:kirtivi. Adella a:negaḷu oḷagaḍe bandu tinda:kuḍte. Hinde ella ja:sti o:pan ittu, all-alli beṇḍe mara ja:sti ittu, adu tintittu. Kaũri mara ja:sti ittu, adu tintittu. A:mele ba:ṇe hullu ittu, adu tintittu. A:mele i: beṭṭa solpa o:pan agidella, me:le, a: opan ja:gake oitiddõ, oḷḷe me:vu siktittu alli, siguru hullu anta. Ondu sari benki bidda:ga siguru hullu bartittu. A: siguru hullu oḷḷe me:vu irtittu. A:mele i:ga avella sigalla adakke. Ya:kandare beṭṭa me:le pu:ra la:ṇṭa:na a:goitu. I:ga ba:ṇe hullu, beṇḍe mara, kaũri mara, ella kaḍame a:goitu.*
> Twenty years ago, the forest was more open. Back then, the gaur, barking deer and elephant all had good food to eat. It's for that reason that they never troubled humans, or came to our fields. Nowadays there's lots of *Lantana*. That's why they have no food. There's open space around the fields, right? Such places are good, and they can easily go in. The elephants go to the places where we plant jackfruit and banana, and eat it all. The forest was more open earlier, and there were more *beṇḍe* (*Kydia calycina*) trees, *kaũri* trees (*Helicteres isora*) and *ba:ṇe* grass. The elephants would eat all of these. The hills used to have open spaces on top, and the elephants would go there. There'd be good fodder for them there, namely grass shoots. The grass would sprout anew after a fire, and those grass shoots were their favourite food. The elephants can't find those foods anymore, because the hill-tops are all covered in *Lantana*. Nowadays it's hard to find *beṇḍe* trees, *kaũri* trees and *ba:ṇe* grass.

6.5 Conclusion

In this chapter, I have showcased the Solega's understanding of seasonal meteorological and ecological cycles, which have a strong bearing on the availability of wild forest products, and on the Solega's agricultural and religious practices. The ecological interactions discussed above include not only anthropocentric relationships of an immediately practical nature, but also third-party commensalisms and antagonisms that highlight the Solega's highly empathic attitude towards the forest's other inhabitants.

Solega people clearly regard many forest animals as undergoing the same tasks and facing the same challenges as they themselves undergo and face in their daily lives. After all, animals are born, get sick, grow old and die in much the same way as humans do. Having themselves experienced the hunger brought on by unfavourable climatic conditions, or through the local extinction of food plants, the Solega are quick to recognise the plight of other forest creatures who are forced to venture ever closer, and more frequently, to human settlements and fields, simply to find suitable grazing. Seeing their forest paths overgrown with an impassable wall of thorny *Lantana*, they understand that it must be equally difficult for the forest's larger animals to migrate to fresh pastures or breeding grounds. Keenly missing the heady scent of once-common fragrant blossoms, they find it unsurprising that fewer beehives have been seen in recent years. And as they watch their immediate surroundings being smothered under a deceptively lush monoculture, they realise that the forest itself is dying, slowly choking to death.

Chapter 7
Honeybee Lore

7.1 On the Manipulation of Resources

Much has been written, in recent decades, on the biological classificatory systems of non-industrial language communities. The communities studied to date are quite varied in their mode of subsistence, and include purely hunter-gatherer and purely agrarian societies, as well as those that practice both to some extent. While detailed cross-linguistic comparisons of folk taxonomies are rare, the few that have been carried out suggest some regular trends for a few variables. In his pioneering study, Brown [113] compared close to 40 languages for which 'reasonable data' were available, and found that languages that have over 330 plant and/or 420 labeled animal taxa tended to be spoken by small-scale agrarian societies. In contrast, the languages of hunter-gatherer societies showed a strong tendency to possess a far smaller number of plant and animal names. Brown attributes this difference to a variety of causes, including a newly-developed interest in domesticated organisms and an increased reliance on 'famine foods' by agriculturalists. Brown also suggests that the actions of agriculturalists might lead to an increase in habitat types, and hence local biodiversity.

Berlin [9] continued this theme in *Ethnobiological Classification*, citing earlier studies which showed that 'cultivated' plant species tend to have far greater numbers of specific-level subdivisions of generic taxa, followed by plants that are considered 'significant', 'protected' and finally 'not treated'. In analysing data from his own studies on Tzeltal and Aguaruna, Berlin defines 'cultivated' species as those that are "deliberately planted and managed by constant and direct human intervention" (p. 120). In contrast, 'protected' plants are those "not consciously destroyed in normal horticultural activities", 'significant' plants are those "that yield useful products but are not systematically protected", while the remainder, 'not treated', are associated

A. Si, *The Traditional Ecological Knowledge of the Solega*, Ethnobiology,
DOI 10.1007/978-3-319-24681-9_7

with "no reported cultural utility". Suggesting a possible mechanism to explain why agriculture and nomenclatural proliferation should be so linked, Berlin states:

> There seems to be little doubt that domestication led, and continues to lead, to the creation of folk specific taxa. It also follows that a cognitively qualitative difference manifests itself as part of the process of human beings' conscious construction and manipulation of new and perceptually different forms of life. Based on this qualitative difference in their interaction with living things—a kind of interaction that is in many respects consciously experimental—individuals begin to take in what might be called a second, more careful look at nature. People begin to be more systematic in the way they deal with the biological world. Regions of the biological space that they had known all along as undifferentiated generic gestalten are now looked upon in greater detail, perhaps even submitted to *close study* [Berlin's emphasis]. Heretofore unnoticed objective differences are recognized explicitly for the first time. Two closely related but unimportant species, at one time unremarkable from a cognitive point of view, now become worthy of linguistic recognition—as distinct parts of what was once a single folk generic taxon. (p. 286)

The generalisations in the quote above give the reader the impression that the "cognitively qualitative difference", which occurred in the minds of agriculturalists, and led them to coin binomial folk-specific names for various crop plants could also lead to a greater differentiation of "generic gestalten" in other biological domains as well—birds, fish and mammals, perhaps. However, Berlin restricts his discussion to the topic of agriculturally significant plants, making no mention of how this proposed mechanism might apply to various animal taxa in the same culture. Is Berlin suggesting that the expertise gained in making more fine-grained distinctions between crop varieties is somehow transferred to other biological domains? Psychological experiments indicate that this is an unlikely scenario [210]. More troubling is his very strong claim that people whose languages contain folk-specific lexemes (keeping in mind that folk-specifics are indicated by their binomial nature) have a "more careful", "more systematic" view of nature, one possessing "greater detail", as a result of possible "close study". At the same time, Berlin implies that certain "objective differences" probably go "unnoticed" by hunter-gatherers, the only evidence being that they do not have appropriate labels for subgeneric taxa. I discuss these nomenclatural issues in some detail in Chap. 2 (see especially Sect. 2.6 for a discussion on the issue of folk-specifics in Anindilyakwa and other Australian languages). In this chapter, however, I will restrict my discussion to the issue of the depth of knowledge that hunter-gatherers might have about an organism that, among the Solega, is not "managed by constant and direct human intervention"—the honeybee.

Although the Solega can be classed as 'agriculturalists', by virtue of their cultivation of finger millet, corn, pumpkin, banana and more recently, coffee, they have also traditionally performed subsistence activities that are unambiguously of the 'hunter-gatherer' variety. This includes the collection of a range of foodstuffs (fruit, mushrooms, honey, edible leaves, etc.), medicines, cleaning and personal grooming products, construction materials and materials for clothing from wild plants which grow naturally in a range of forest habitats (Sect. 3.9). In the following, I will argue that the Solega's knowledge of the biology of honeybees (a resource that they 'gather' from, rather than 'constantly manage') is remarkably accurate and consistent, given the types of observations that they might have the opportunity to make during their interactions with these insects. Admittedly, neither Brown nor Berlin has explicitly made any mention of differences in the depth or—if I might use a provocative term—quality of the traditional biological knowledge of the speakers of different languages. However, as the passage quoted above suggests, certain vague, and in my opinion, unjustified, assumptions regarding the pre-eminence of the agriculturalist way of life underpin Berlin's ideas on the "evolution of ethnobiological categories", where the emergence of subgeneric taxa is taken to be a sign of increasing complexity in an ethnoclassification system.

> Plants and animals evolve over time. The systems of ethnobiological classification that human beings come to develop to mirror evolution's work also change through time... (p. 272)

Berlin never clarifies the exact relationship between biological evolution and his own conception of evolving ethnoclassification systems, but it is clear that the analogy alluded to in the above quote cannot be taken literally, due to the vast differences in the temporal scales over which the two might operate. Moreover, his idea of an evolutionary sequence that involves an increase in 'complexity' highlights some worrying assumptions implicit in the enterprise of comparing classification systems across languages. Such assumptions have long plagued western thought concerning the natural world, and, in the most famous instance, have taken the form of the now outdated and highly anthropocentric *Scala Naturae* 'the ladder of nature', where Man was placed at the top of all earthly creations, superseded only by God and His angels. While this dogma is no longer a part of western scientific thought in a literal sense, there are indications that such ideologies still permeate much of modern evolutionary biology. Many biologists unconsciously use the word 'complexity' to describe the random directionality evident in many evolutionary events, without considering the broader implications.

McShea [211] has argued that even though modern-day multicellular organisms are indeed more complex than the bacteria-like organisms that comprised the earliest life forms on the planet, there is no hard empirical evidence to suggest that increasing complexity is a guaranteed, or even frequent, outcome of long-term evolutionary processes. Advocating an "emphatic agnosticism" on the question of increasing complexity, McShea wonders "*whether the word "complexity" (as it is commonly used) is just a modern substitute* [for the *Scala Naturae*], *a kind of code word for perfection, progress, and proximity to us*." His main objections to the exercise of comparing species separated by long stretches of evolutionary time are not only that 'complexity' is a vague and ill-defined term, but also that there can be several (often conflicting) metrics of complexity—number of genes, size of the genome, body size, number of limbs, relative brain size, specialization of body segments, degree of brain folding, just to name a few—and that the ad hoc choice of any one of them might provide an appealing, but inaccurate, answer.

An even stronger argument was made by the evolutionary biologist Stephen Jay Gould, who devoted an entire monograph to the question of anthropocentrism embedded in the "poorly formulated and improbable" ideas of progress in long-term evolution [212]. Gould notes that such ideas are so compelling, and so prevalent in modern evolutionary thinking (numerous textbooks illustrate the idea of evolution with the clichéd fish–amphibian–reptile–land mammal–human sequence) that even experienced academics come to take such ideas for granted, without considering vital issues such as evidence and causality. In the second chapter of his book (p. 28), Gould offers a quote from *The Diversity of Life* by E. O. Wilson, for whom he has a great deal of admiration. Wilson also appears to conflate evolution with progress, and defends his view with a final line that Gould says he "*almost found chilling*":

> During the past billion years, animals as a whole evolved upward in body size, feeding and defensive techniques, brain and behavioral complexity, social organization, and precision of environmental control... Progress, then, is a property of the evolution of life as a whole by almost any conceivable intuitive standard... let us not pretend to deny in our philosophy what we know in our hearts to be true.

Gould goes on to suggest a mechanism by which a trend of increasing complexity can appear out of processes that are directionally random. Simply put, bacteria are the simplest forms of life possible, and further simplification is prevented by an evolutionary 'wall'. The result is that even random evolutionary events will necessarily produce more complex organisms over time, thereby giving the illusion that the process of evolution is an inherently complexity-increasing one. Gould also liberally cites the work of McShea—in particular, his point that it is important to distinguish 'passive trends' (overall results arising as incidental consequences, with no favoured direction for individual species) from 'driven trends' (each species of a lineage tends to change in a particular manner, because evolution supposedly favours more complex creatures).

Applying these ideas to ethnoclassification systems, it is tempting to arrive at the same conclusion as Berlin—that hunter-gatherer languages that lack folk specifics (and are instead perhaps dominated by folk generics with little explicit, higher order grouping) represent the 'simplest' possible ethnoclassification systems, and that the advent of agriculture resulted in increased 'complexity' simply because there was no other way to go. However, one crucial difference between the biological and the linguistic data is the time-depth of the available information. While biologists can rely on an imperfect, but temporally extensive fossil record (as well as extant microorganisms that are commonly believed to be good exemplars of the earliest forms of life to have evolved), ethnobiologists usually only have access to synchronic linguistic (and cultural) data that provide little evidence of how current systems may have come into being. Of course, there exist landmark studies that have made use of empirical, diachronic data from complementary fields such as linguistics, archaeology and genetics in order to describe the evolution of cultural phenomena such as subsections, migration and exogamy [213–215]. Unfortunately, comparable studies do not appear to exist in ethnobiology, with the result that strong claims about the evolutionary direction of ethnoclassifications are made solely on the basis of synchronic data.

Similar objections have been raised in relation to the idea of Cultural Evolution, that appeared in the anthropological discourse of the nineteenth and mid-twentieth centuries in two distinct guises [216]. Using contemporary European society and cultural 'development' as a benchmark for the human species, proponents of the former phase assigned other cultures to three developmental stages—savagery, barbarism and civilised—and theorised that this was the trajectory taken by all societies. In the second, more modern instance of Cultural Evolution, the focus shifted from the crude generalisations of the old system to what was thought to be a set of empirically sound, scalable traits. Here, too, a developmental sequence was proposed: band, tribe, chiefdom, state. In his critique of both schools of thought, Dunnell [201] points out how the former struggled to come to terms with the continued existence of traits categorised as primitive in many societies around the world, and with the lack of any apparent 'development' (comparable to Europe) in the archeological record of Australia and North America. Supporters of the second phase, who attempted to order societies using scalable empirical traits, were guilty of choosing traits that were, "*by and large, traditional* ad hoc *descriptive categories*" (p. 178). Dunnell concludes that for such academics, "*increasing complexity or progress is not discovered by scaling societies, but rather it is built in a priori by trait selection*" (p. 178).

Berlin admits that some anthropologists might object to his use of the word 'evolution' (p. 272), but the above discussion makes it clear to see that at least some prominent evolutionary biologists share the anthropologists' misgivings. The existence of subgeneric folk taxa in the classificatory systems of some languages might indicate a higher level of 'morphological hierarchical complexity'—just one of McShea's four metrics—over languages that lack this level *sensu* Berlin [9]. However, to conclude that the former type of system is more complex than the latter is, in my opinion, premature—the use of a metric other than the presence or absence of binominal subgeneric taxa might reveal a very different picture. To hypothesise that the taking up of agriculture by previously hunter-gatherer populations is usually accompanied by a proliferation of subgeneric taxa in their language (i.e. a historical correlation) seems unproblematic, especially if a mere 'passive trend' is being posited. It is quite another matter, however, to posit a causal relationship, and claim that agriculturalists take a "more careful" look at nature, and that this high level of scrutiny allows them to detect hitherto unobserved patterns within the "undifferentiated generic gestalten" that they had previously taken for granted. The comprehensive list of Anindilyakwa plant and animal names complied by Waddy [26] is meant to be in keeping with Berlin's views on the ethnoclassification systems of hunter-gatherer communities—it completely lacks any subgeneric taxa, and so must be of a lower complexity than the classificatory systems of agricultural communities. However, as Baker [13] has cogently argued, Australian languages are, in general, able to function just as well as the languages of agriculturalists in the absence of binomial subgeneric taxa, with their speakers making fine-grained distinctions between hundreds of closely-related biological species.

The Solega cannot be satisfactorily placed within the neat agriculturalist/hunter-gatherer dichotomy, as they not only cultivate food crops such as finger millet and corn, but also forage for an extensive range or forest products for daily use. As mentioned earlier, they do not keep honeybees, but instead collect wild honey in an organised or opportunistic fashion. I hope to argue in this chapter that the Solega's understanding of the natural history of honeybees—exploited, but not managed, wild organisms—is at least as good as that of societies that have indulged in beekeeping for centuries. In doing so, I hope to provide evidence to counter the claims implicit in Berlin's hypothesised mechanism of the 'evolution' of classification systems.

7.2 The Word *je:nu*

The Solega word *je:nu* is polysemous, and can mean honey, honeybee (an individual insect) or even a honeybee colony or honeycomb. In Kannada, although the morpheme *je:nu* occurs in the words 'honey' (*je:nu tuppa*) and 'honeybee' (*je:nu noṇa*), one must necessarily make use of the full lexemes, in order to distinguish between the insect and its product. Solega speakers, however, often refer to the individual insect, the colony/comb, as well as the harvested product, using just the word *je:nu*, as exemplified below:

***je:nu* = individual bee(s)**

kembare.bare	*je:nu*	*ken-je:nu*
place.name	bee	red-bee

'The bees found at Kembare Rock are red bees.' (excerpt from a song)

a:ga:	*ho:gi*	*fffff...*	*a:ge*	*ke:ḷ-a*	*be:ku*	*oḷa:ge*
then	go.CONV$_2$	(puff.of.air)	thus	listen-CONV$_3$	need	inside

bush	*en-d-a:de*	*je:nu*
ONOM	say-CONT-3SG.N	bee

'Then you go and (blow into the hole), you need to listen like this, the bees inside go 'BUSH'.'

***je:nu* = honeybee colony/comb**

nalavattu	*aivattu*	*je:nu*	*band-a:de*	*ond-e*	*mara-ka*
forty	fifty	colony	come-3Nsg	one-EXCL	tree-DAT

'Forty or fifty honeybee colonies come to a single tree.'

Ondu	*mara*	*idda-le*	*a:*	*mara-ka*	*je:nu*	*ku:t-ur-de*
One	tree	be-COND	that	tree-DAT	colony	sit-IMPERF-3SG.N

'If there's a tree, a colony will come and sit on the tree.'

iṭṭu	*koda:ṇi*	*je:nu*
pollen	bearing	comb

'pollen-bearing comb'

***je:nu* = honey**

je:n-iga	*ele kitturu*	*ba-ḍau*	*kunna*
honey-DAT	leaf pick.CONV	come.IMP-ADD.NONHON	boy

'Pick some leaves for the honey and come here, boy!' (leaves for a receptacle)

a:	*maruḷi*	*hu:-v.ina*	*je:nu*	*baṇḍ-a:de*
DIST	tree.name	flower-GEN	honey	good-3Nsg

'Honey from that maruli flower is good.'

One consequence of this polysemy is that for some uses of *je:nu*, the English translation can be quite ambiguous. However, this is hardly ever a source of confusion for the Solega listener, for the simple fact that honey, honeybees and honeybee colonies/combs tend to co-occur in space and time:

aravilu.kende	*mara.d-alli*	*je:n-a:de*
tree.name	tree-LOC	honey(bee/comb)-3SG.N

'There is/are honey(bees)/(bee colonies) on the aravilu kende tree.'

a:	*ja:ga:.d-alli*	*je:nu*	*sikk-a:d-ille*
DIST	place-LOC	honey(bee/comb)	find-GER-NEG

'You won't find honey(bees)/(bee colonies) in that place.'

7.3 Honey Harvesting

The honey harvesting technique of the Solega is quite straightforward, and in the case of *hejje:nu*, at least, involves the production of smoke. The peak harvesting season for *hejje:nu* is May–June, although there can be up to three main harvests carried out in 1 year. A small pile of twigs (*phuḷḷe*) is set alight, and then covered with fresh leaves (*soppu*), which results in the production of thick smoke. This whole bundle can be tied together to make a smoking torch (*su:ṭe*), which is carried up the tree to target individual hives. Honey from *hejje:nu* colonies is usually harvested at night, as this is said to reduce the number of stings received by the humans involved in the harvesting operation. This is particularly true when harvesting honey from the 'bee trees', which, as mentioned above, may be home to several dozen large colonies. Daylight collection of honey is possible for the other three types of *je:nu*, as well as for hejje:nu colonies occurring singly. One informant explained that night-time harvesting results in fewer stings because bees cannot see very well in the dark.

Collecting honey from a *hejje:nu* colony involves cutting out the honey-bearing 'head' of the colony with a knife or sharp stake (*aṇegaḍḍi*), and leaving the rest of the comb unmolested. The excised honeycomb is lowered to the ground in a basket tied to a rope (*ma:linagga* or *ma:lina naru*) made from the vine *ku:guri ambu*. The honey is separated from the comb, and the wax is kneaded into balls. Back home, the wax balls are dissolved in hot water, sieved through a cloth, and sold in town.

In the case of *thuḍuve je:nu*, the collector inserts his hand into the tree hollow or rock crevice occupied by the bees, and feels as many of the parallel wax combs as possible. An experienced collector can distinguish honey-laden combs by their consistency; these are then preferentially harvested. If the collected honeycomb is found to contain some brood, the honey and brood can be eaten together, as long as the brood is still in the larval stage. Older, pupating (and presumably capped) brood, which has developed wings, is discarded. The larvae are said to contain a milky substance, which is quite tasty in combination with honey.

The harvesting of these two types of honey can therefore be labeled 'sustainable', as much of the hive and brood is left intact, and the bees are usually able to rebuild the missing comb, and collect new food stores. Moreover, when harvesting from 'bee trees', a few colonies are always left untouched, to allow the population to build up again. For this reason, it is possible to harvest honey up to three times in a year from the same *hejje:nu* colony. In contrast, the collection of honey from *kaḍḍi je:nu* and *nesari je:nu* hives tends to result in the destruction of the hive. For the former, since the top of the wax comb wraps around the supporting branch, it is usually severed towards its proximal end (with respect to the central trunk of the tree), and the comb, with the branch running through it, is gathered whole. In the latter case, the small size of the nest makes it difficult to selectively harvest some parts of the comb, while preserving others.

Solega honey harvesters, who have been involved in the sale of honey to non-Solega tradespeople, are able to give estimates of typical honey yields from each type of bee. A full *hejje:nu* comb could hold as much as 25 kg of honey, a *thuḍuve je:nu* colony could store 5–7 kg, while a kaḍḍi *je:nu* colony might yield 0.5–1 kg. *Nesari je:nu* hives are usually raided for an on-the-spot meal, but might also hold the same amount of honey as a *kaḍḍi je:nu* colony.

Excursus 7.1: A Beesting Remedy

No action is taken when one receives a few stings during the course of honey-gathering activities. When something goes wrong, however, and the gatherer is attacked by *en masse*, the following remedy can be employed:

> *Namage ja:sti je:nu kaccu uṭṭre, nammalli a:sa:ḍi be:ru anta ondu giḍa ide. Adava na:vu tegedu asi:ne tintivi. Adu je:nu kaccidare ma:tra. Geṇasu tara irutte. Me:le tokke irutte, tokkege tegedu bisha:ku uṭṭu oḷagaḷa ba:ga tintivi. A:va:ga visa ella eḍdu ho:gu uḍte, no:vu solpa beṇḍa:gotte.*
>
> We have a plant called *a:sa:ḍi be:ru* for when we get stung a lot. We look for it, and eat it raw, but only when we get stung by bees. It's like a yam. There's bark outside, we remove the bark, throw it away, and eat the inner part. Then the poison goes away, and the pain lessens.

7.4 Bee Songs

Like the Jenu Kurumba and the Alu Kurumba of the nearby Nilgiri Hills, the Solega also have a 'honey song', which is sung prior to, or during, a honey-harvesting event. Unlike the highly elaborate song of the Kurumbas, however, the Solega honey song is short and simple, although it can be sung with different melodies, and with a few textual variations. Following is one of three versions of the Solega honey song recorded in April 2009 from native speakers in three settlements:

aṇṇa:ne: ka:ḍiga na:daru neḍeyaṇṇo *aṇṇa:ne: ka:ḍiga na:daru neḍeyaṇṇo*	Brother, let us go to the forest, Brother, let us go to the forest.
aṇṇa:ne: kembare bareyalli je:na:de *aṇṇa:ne: kembare bareyalli je:na:de*	Brother, there is honey at *Kembare Bare*, Brother, there is honey at *Kembare Bare*.
aṇṇa:ne: mu:ru moḷada muṇḍu su:ṭe *aṇṇa:ne: a:ru moḷada aṇegaḍḍi*	Brother, a three-*mola*[1]-long *muṇḍu su:ṭe*,[2] Brother, a six-*mola*-long *aṇegaḍḍi*,[3]
aṇṇa:nu: ka:ḍiga na:daru neḍeyaṇṇo *aṇṇa:ne: ka:ḍiga na:daru neḍeyaṇṇo*	Brother, let us go to the forest, Brother, let us go to the forest.
aṇṇa:ne: kammuri bareyalli je:na:de *aṇṇa:ne: kammuri bareyalli je:na:de*	Brother, there is honey at *Kammuri Bare*, Brother, there is honey at *Kammuri Bare*,
aṇṇa:ne: no:ḍi kuiyalo: je:nanna: *aṇṇa:ne: no:ḍi kuiyalo: je:nanna:*	Brother, look, let's harvest the honey, Brother, look, let's harvest the honey,
aṇṇa:ne: ka:ḍiga na:daru neḍeyaṇṇo *aṇṇa:ne: ka:ḍiga na:daru neḍeyaṇṇo*	Brother, let us go to the forest, Brother, let us go to the forest.
aṇṇa:ne: no:ḍi cuccalo: je:nanna *aṇṇa:ne: no:ḍi cuccalo: je:nanna*	Brother, look, let's poke out the honeycomb, Brother, look, let's poke out the honeycomb,
aṇṇa:ne: ku:guri (h)ambina kuḍima:lo: *aṇṇa:ne: ku:guri (h)ambina kuḍima:lo:*	Brother, a rope made of *ku:guri* vine, Brother, a rope made of *ku:guri* vine.

The song describes in some detail the implements used during the harvesting process, as well as the actions performed; in places, the singer also mentions locations where a particular kind of bee is to be readily found. The place *Kembare Bare* is also mentioned in other variants of this song, which indicates that this location holds special religious and/or cultural significance for the Solega:

kembare bareyalli je:na:de *ka:rana kareyo ka:raĩyya*	There is honey at *Kembare Bare*, (Invocation of the *God Karaĩya*)
kembare bare je:nu kenje:nu *ka:rana kareyo ka:raĩyya*	The bees at Kembare rock are red bees, (Invocation of the *God Karaĩya*)
soppi nerake ka:ra:na	O *Ka:ra:na* who dwells in the shade of the leaves!

Sometimes, other honey-gathering locations are mentioned, as in the following:

haddu bare je:nu adugidade *soppi nerake ka:ra:na*	The bees at *Haddu* rock sway in the wind, O *Ka:ra:na* who dwells in the shade of the leaves!

and:

a:la bareli je:na:de *na:ḍe: ni: nu:r ondu je:nu adeyaṇṇa*	There are bees at *A:la* rock, There are a hundred and one hives in the forest.

[1] A measure of length, taken to be the distance between the elbow and the middle fingertip.

[2] A torch used for making smoke.

[3] An implement for cutting out honeycomb (K: *pacce*).

Other place names such as *naku bare*, *hoṇṇe bare* and *du:pa bare* are also periodically mentioned. The implements mentioned in the songs may also vary, as in the following:

hotta:ganambina odema:lu *soppi nerake ka:ra:na*	An *odema:lu*[4] made from *hottaga* vine, O *Ka:ra:na* who dwells in the shade of the leaves!

and:

karikamma jo:tura jo:ḷige *aṇṇa:ne:*	A *jo:lige*[5] made from black cloth, O brother!

The content of the Solega song(s) therefore contrasts sharply with that of the Jenu- and Alu Kurumba honey songs. In the case of the former, Demmer [217] writes that:

> …the honey gatherers construct a context for their actions with [a] song which rhetorically creates a common identity with the bees as affinal relative. The gatherers regard the bees as potential marriage partners, and in analogy address the mother of the bees as mother of their affinal partners.

Similar observations were made of the Alu Kurumba honey song by Kapp [218]:

> *Dieses in seiner Melodie von Melancholie getragene und in seinem Wortlaut poesievolle Lied ist seinem Inhalt nach ein von einem jungen verliebten Mann an seine Braut gesungenes Liebeslied.*
> This song, with its melancholy melody and its poetic wording is, as revealed by its content, a love song, sung by a young lovesick man to his bride.

To my knowledge, the kind of honey song described for the Kurumbas does not exist in the Solega oral tradition. While there are reports that one Solega honey song mentions "how the bees collect pollen from different flowers" [88], there seems to be no attempt to placate angry bees by claiming affinal relationships with them. Affinal relations such as *aṇṇa* (brother) are indeed mentioned, as is *ba:maida* (brother-in-law), as in the following:

paṭi hoṭi enda:de ba:maida *soppi nerake ka:ra:na*	(The rope) snaps, O brother-in-law, O *Ka:rana* who dwells in the shade of the leaves!

These, however, are most likely directed at other members of the honey-gathering group, who may indeed by relatives of the singer. In any case, the Solega's awareness of honeybee biology rules out the possibility that they might be referring to the bees as 'brother' or 'brother-in-law' (see below).

[4] Another kind of rope.

[5] A folded cloth purse for carrying honey.

7.5 Solega Knowledge of Bee Natural History

Honeybees are plentiful in the B. R. Hills, and the Solega readily exploit this resource during particular times of the year. During the months from March to July, honeybees from the lowlands migrate into highland forest areas to take advantage of the seasonal flowering of large rainforest trees. It is also during this time that the Solega start to be keenly aware of the presence of bees in their environment, frequently looking up at trees to look for hives, and exchanging information on the movements of bee colonies in the neighbouring forest.

7.5.1 *Types of Honeybees*

The Solega recognize and name four types of bees (see Fig. 2.5). The four types of *je:nu* in Solega are *hejje:nu* (*Apis dorsata*), *t(h)uḍuve je:nu* (*Apis cerana*), *kaḍḍi je:nu* (*Apis florea*) and *nesari je:nu* (*Trigona iridipennis*). There is said to be a 'kinship' relation between the different bee species, and as far as I know, this is the only example in Solega of natural kinds being regarded as kin to one another. This phenomenon is further described in Excursus 7.2 (see below), but briefly, it is probably a 'just-so' story to explain why the various bee species do not visit each other's hives. The term *je:nu*, as applied to the insect, differs markedly from the western biological concept 'bee', being used by the Solega to refer to only those bee species from which honey is harvested. Many other species of solitary bee are to be found in BRT, including large carpenter bees and the colourful blue-banded bee, *Amegilla* sp. These organisms are unanimously labelled *dumbi*, a category whose prototypical forms are large beetles. In addition to the four main 'subgeneric' categories of honeybee, the Solega name what can only be described as two types of 'varietals'—*ko:ḷi je:nu* and *kenje:nu*. The former is a type of *kaḍḍi je:nu*, and is so named because its hives are found closer to the ground. The latter is a type of *tuḍuve je:nu*, and the special name refers to the reddish colour of these bees.

The Solega honeybee taxonomy presented here shows many deviations from Berlin's predictions. First, the Solega 'generic' taxon *je:nu* does not exactly overlap with either the biological genus *Apis* (as the stingless *Trigona* are also included in the former category), or with the English[6] label 'bee' (as solitary and carpenter bees are excluded). Second, there are four distinct 'subgeneric' binomial labels for the four honeybee species, which is at odds with the fact that as far as honey is concerned, the Solega are very much 'gatherers' and not 'beekeepers'.

[6] Species such as the blue-banded bee are definitely called 'bees' in non-technical English—photographs from the websites of amateur photographers often label *Amegilla* sp. simply as "blue bee". Several wasps are also frequently labeled as "bee" on these websites, showing that (a) the photographers are usually not trained biologists, and (b) the semantic ranges of the English and scientific 'bee' are quite different.

A staunch supporter of Berlin might argue here that as the Solega cultivate *ra:gi* 'millet' they should really be classed as agriculturalists, who would be expected to have subgeneric taxa. The reality is that the Solega do both, and until some causal effect of agriculture on the perception of other biological entities can be demonstrated (or even a plausible mechanism described), it makes no sense to assume that the cultivation of a food crop can somehow cause people to better distinguish between honeybee species. Ellen [72] has reviewed the evidence used by Brown [113] and Berlin [9] to make their case for a correlation between mode of subsistence and the ethnobiological lexicon. Ellen was able to identify a number of confounding factors in both data sets, including the fact that the 'hunter-gatherers' included in the comparisons were from temperate and subarctic latitudes (where biodiversity is generally lower); by the same token, the counts of generic terms were higher for both hunter-gatherers and agriculturalists in tropical rain forests.

The most parsimonious explanation for the existence of subgeneric honeybee categories in Solega appears to be a utilitarian one—simply that the four species are distinguished and named because they are exploited by the Solega. Finally, the 'varietal' names *ko:ḷi je:nu* and *kenje:nu* have the same structure as the taxa of the 'subgeneric' level, and do not incorporate the names of the superordinate category. Structurally, then, *ko:ḷi je:nu* and *kenje:nu* appear to form a contrast set with the four main honeybee species, but Solega speakers will explicitly state that *ko:ḷi je:nu kaḍḍi je:nige se:rutte* '*ko:ḷi je:nu* meets/belongs to *kaḍḍi je:nu*'. Biologically, too, *ko:ḷi je:nu* and *kaḍḍi je:nu* belong to the same species, but may well represent separate subspecies. Similarly, *kenje:nu* is said to be a kind of *tuḍuve je:nu.*

All the Solega speakers interviewed in the field were able to provide detailed information on various aspects of the biology of the four bee species: this included their life histories, hive architecture, migration schedules, preferred nesting sites, and times of high honey flow. Most were also able to identify the tree species on which certain kinds of bee were most likely to be found. The same tree species were repeatedly named by various speakers, indicating a high level of precision, and therefore a very real (biological) preference among bee species for certain kinds of trees.

7.5.1.1 Hejje:nu

The name *hejje:nu*, (in particular, the presence of the prefix *heʔ-*) acknowledges the fact that this is the largest type of honeybee known to the Solega. The colonies of *hejje:nu* (the giant honeybee or *Apis dorsata*) are common in high-altitude rain forests (*male ka:ḍu* or *ka:nu ka:ḍu*), although they may also be found in dryer, lowland forest types (*na:ḍu ka:ḍu*). Place names where *hejje:nu* may be found are readily volunteered: *Ambaḷḷa* (the name of a stream, or *aḷḷa*), *Doḍḍa Sampage Ka:ḍu* (the forest surrounding the sacred *doḍḍa sampage tree*), *Seppirotte Ka:ḍu*, *Aravilu Ka:nu* (an evergreen forest dominated by *aravilu* trees), *Daiya Gallu* (a rocky place which is home to a *daiya*, a malevolent female spirit), *Kembarre* (a rocky outcrop or arre named after the *kemba* trees growing near it) and *Thaṇiki*.

In evergreen forests, *hejje:nu* prefer to live on very tall trees, showing a marked preference for the *soravilu* (*Acrocarpus fraxinifolius*), *ba:ge* and *sele* (*Albizzia odoratissima*, *Albizzia lebbek*) trees. These two names were repeatedly offered by Solega consultants from a number of widely-dispersed settlements, which suggests that *Apis dorsata* colonies do indeed favour these species as potential nest sites. Indeed, individual trees of these species may be well known across a range of Solega settlements as *je:nu mara* or 'honey trees', due to the fact that they are home to a large number of hejje:nu colonies year after year. For instance, the *Do:vu Ma:vu Ba:ge* is a single large *Albizzia odoratissima* found near *Ko:li Ba:vi* hill, which attracts upto 50 *hejje:nu* colonies around the same time every year. An invitation to go to *Do:vu Ma:vu Ba:ge* at the right time of year can only mean a honey-harvesting trip; no further explanation is required. Interestingly, this particular tree derives its name from the locality where it is situated, which, in turn, derives its name from other trees that used to grow there in the past. The *ma:vu* of *Do:u Ma:vu Ba:ge* indicates that a group of mango trees, presumably another important food source, grew at this a spot (which, incidentally, lies next to a stream). Thus, the location, which is itself identified by the (former) presence of a particular kind of tree, is now used to pinpoint the location of an individual of another tree species. Several other 'bee trees' are known to the Solega (Sect. 3.8.2), and these, along with other locations where bees often nest, are remembered a mental map of honey-harvesting sites. In the following passages, a single consultant (MRM) is asked to recall the important honey-harvesting sites known to him, and the directions for getting to these places. His monologue is transcribed here in its entirety because of the highly detailed nature of his recall:

Ro:ḍe: ho:ga:du keḷeka, basappana kaḍaviga. Alli aranelli mara ade. Aranelli maradalli ka:vala ha:kidivĩ. Je:nu ondu-eraḍu tinga ka:tivẽ. Eraḍu tinga ka:takki a:mele je:na kuidivẽ. Adu doḍḍa koḷḷada aranelli... A:mele ha:geve keḷeka bandale, madyadalli de:varu ade—basappã. Basappana keḷeka bandale beccarre ade. Beccarru bareli uwe je:na kuidivẽ. Ade aḷḷave keḷeka bandale neḍeduru neḍeduru, keḷage iḍi haḷḷi bare ade. Iḍi haḷḷi bareli je:na kuidivẽ.

You need to go downhill along the road, to *Basapanna Kaḍavu*. There's an *aranelli* tree (*Garuga pinnata*) there. We keep watch at the *aranelli* tree. We wait one or two months for the bees. After a couple of months, we harvest the honey. That's the *Doḍḍa Koḷḷada* [big valley] *aranelli*... Then if you keep going down, there's a holy site in the middle—[the god] *Basappã*. *Beccarre* is downhill from there. We also harvest honey from *Beccarre* rock. If you keep walking downhill along the stream, you get to *Iḍi Haḷḷi* rock. We collect honey from *Iḍi Haḷḷi* rock as well.

Doḍḍa sampageyinda solpa munde ho:dale soravilu maradalli je:nu ade. Na:ku soravilu—na:ku soravilinalli ondu eraḍu soravilu satto:tu, innondu eraḍu soravilu ade. Innu ha:geve bandale mele guṇḍu sikkida abbili alli ondu soravilu mara ade. Adakka ondu hattu je:nu banda:de. Innu ha:geve me:leka me:leka bandale alliga gombegallu bantu. Gombegallinda me:leka bandale alli sikka sampage ade. Sikka sampage me:laḷa aravilu beṭṭa ade. Alli ondu aḷḷa banda:de. Doḍḍa sampage aḷḷaka ue sikka sampage aḷḷaka uve madya se:ridde adu. Sikka sampage aḷḷave me:leka ho:dale alli ondu aravilu kende ade. Adakka ondu ippattu je:nu banda:de. Ada kuiduru bandivĩ. Aravilu beṭṭada keḷageve irada mara. Ondu arre me:le ade a: mara. A:mele sikka sampage me:leka ho:ku. Me:leka ho:da:ge alli ondu tekke soravilu mara ade. A: tekke soravinalli ondu ippattu je:nu banda:de. Ada kuiduru solpa balaka turugidale alli iṭṭu bu:di ade. Iṭṭu bu:dili eraḍu mu:ru soravilu ade. Adakka je:nu banda:de. Ada kuiduru bandivĩ. Iṭṭu bu:dinda kuiduru me:laḷa ro:ḍiga ho:dale kambaḷi gadde ka:niga, alli jo:ḍu turuve ade. A: jo:ḍu turuve maradalli je:nu kuidivĩ. Jo:ḍu turuve endale ro:ḍu madya ade, a:se ondu mara, i:se onde mara ade. Adu eraḍu mara ira kaṇḍu na:ve jo:ḍu turuve endivĩ. Doḍḍa doḍḍa mara—madyada me:le se:ru uṭṭade eraḍu uwe. I: jo:ḍu turuveyinda a: jo:ḍu turuveka kaṇeya:na maḍaga:ku.

If you keep going past *Doḍḍa Sampage*, there's a *soravilu* tree with bees. [There were] four *soravilu* (*Acrocarpus fraxinifolius*) trees [initially]—one or two have died, and there are still a couple left. If you keep going, you'll see another *soravilu* tree at *Guṇḍu Sikkida* waterfall. Ten or so bee colonies nest there. If you keep going uphill from there, you get to *Gombegallu* village. If you climb uphill from there, you'll find *Sikka Sampage* (tree). Above *Sikka Sampage* is *Aravilu* hill. A stream flows from there. It meets both the *Doḍḍa Sampage* and *Sikka Sampage* streams in the middle. If you keep going upstream of *Sikka Sampage*, you'll find an *aravilu kende* tree (?). About twenty colonies come to that tree. We harvest from there. It's a tree that's right below *Aravilu* hill. It's growing out of a rock platform. Next, you can go uphill from *Sikka Sampage.* There you will see a *tekke soravilu* tree (?). About twenty bee colonies come to that tree. We harvest from that tree as well. If you walk a bit to the right from that tree, you get to *Iṭṭu Bu:di*—two or three *soravilu* trees grow there too. Bees nest there. We harvest from those trees as well. When you go up to the road from there, to *Kambaḷi Gadde* forest, you find the *Jo:ḍu* [twin/joint] *Turuve* trees (*Ligustrum perrottetii*). We harvest from the *Jo:ḍu Turuve. Jo:ḍu Turuve* means the road goes in between, there's a tree on one side, and a tree on another. We call those two trees the *Jo:ḍu Turuve*. They're big trees—they meet in the middle [overhead]. You need to build a bridge to climb from one tree to the other.

Keredimba pakkaka a:cari ho:dale si:geguḷi ba:ge alli ira:du, kanneri koloniyinda a:ce. Adakka ondu aivattu nu:ru je:nu banda:de. Alli aivattu nu:ru je:nu kanneri kolonive kuidumadu. Namma kaḍedu alli ho:ga:dille. A:mele modeya:na ta:ri. Alliye—bare beṭṭa, kanneri koloni pakka. Modeya:na ta:rika aivattu je:nu bartittu, munde. A:mele bare beṭṭa. Bare beṭṭaka mu:vattu je:nu bartittu. Bare beṭṭa kuiduru alugaṇi kere nuggi basappana kaḍavina silliga se:ri i: ro:ḍe na:vu je:nu tuppa tartidadu.

If you keep going past *Keredimba* village, you find the *Si:geguḷi Ba:ge* tree, on the other side of *Kanneri* colony. About 50 to 100 trees come to that tree. The people from *Kanneri* colony gather honey from those hives—our people don't go there. Then there's the *Modeya:na Ta:re* tree (*Terminalia bellirica*). It's there too—on *Bare* hill, near *Kanneri* colony. Used to find about fifty hives on the *Modeya:na Ta:ri*. Then there's *Bare* hill itself. About thirty colonies would come to *Bare* hill. We'd collect honey from *Bare* hill, go towards *Alugaṇi* reservoir, arrive at *Basappana Kaḍavu*, and carry our honey along the road.

Other trees species where *hejje:nu* may be found include *ma:vu* (*Mangifera indica*), *alasu* (*Artocarpus heterophyllus*), *kuḍugilu* (*Casearia tomentosa*), *karava:di* (*Persea macrantha*), *karavilu* (?), *honne* (*Pterocarpus marsupium*, specifically during the first rain of the year—*munga:ru*—usually in the month of March), *a:le* (*Holarrhaena antidysenterica*), *matti* (*Terminalia crenulata*), *beḷḷa:ḍe* (*Neolitsea zeylanica*), *ka:nuḍupe* (*Grewia* sp.), *ka:nadaḍasu* (*Grewia tiliifolia*), and *kaggali* (*Acacia chundra*). In the lowland forests, *hejje:nu* may be found on the *ni:ru matti* (*Terminalia arjuna*) tree. It is also common to find *hejje:nu* colonies on boulders (*bare kaṭṭe*).

Hejje:nu start arriving[7] in locations familiar to the Solega in March, at the time when the flowers of the *honne* tree are in bloom. Apart from this very obvious annual pattern of appearance and disappearance that the giant honeybees exhibit, there is some evidence that the Solega are aware of this species' stepwise migration behaviour, which has frequently been reported in the scientific literature. In the following excerpt from an interview recorded in *Bu:ta:ni Po:ḍu*, the speaker contrasts the 'arrival' behaviour of *hejje:nu* and *tuḍuve je:nu* on the one hand, which appear to make more than one stop before deciding on a final hive location, and the stingless bee *nesari je:nu*, which appears to choose a final hive location straight away:

> *Hejje:nu honne hu:vina ṭaim bartade... a:ga hejje:nu be:re be:re ja:ga:ka ho:gi se:rude, ka:ḍe: pu:ra:uve se:rude, tuḍuve ue ka:ḍe: pu:ra:uve se:rude, i: nesari ondu ma:tra yelli ku:titte alliye, alliye.*
> *Hejje:nu* arrives in the time of the *honne* flower... then *hejje:nu* goes to many different places, it goes all over the forest, *thuḍuve* also goes all over the forest, it's only nesari that (remains) exactly where it (first) lands.

As is well known, *hejje:nu* colonies consist of a single large comb, hanging vertically down from a large tree branch or a rock overhang. In Indian cities, it is also not uncommon to see *hejje:nu* combs attached to the eaves of buildings. This bee stores its honey in a discreet band at the top of the comb, leaving the rest of the comb for brood. The part of the comb containing the honey is called *tale* 'head', while the term *ra:ḍe* is used to refer to the brood containing part of the comb, or to the entire comb. Pollen (*iṭṭu*) is also kept in a discreet area, close to the honey store.

Excursus 7.2: Affinal Relations

The four bee species are considered siblings, with the smallest, *nesari je:nu*, said to be the eldest brother:

> *Nesari entivilla, saṇṇadu, avanette iri-aṇṇa. A:ga adakka tamma—hejje:nu. Adakkinta tamma—kaḍḍi je:nu. Adakkinta tamma—tuḍuve je:nu. A:ga gauri habba banda ṭaiminalli ella beka:du beḍa:du tiṇḍi baṇḍi ella se:rsi, "Tamma, ba:ri, nanna meneka, u:ṭakka!" Endu tamma avaru bandu no:ḍdavare—tale nugga:dille, oḷeka. E:kandale ba:gilu sari ille. Tuḍuve, hejje:nu, kaḍḍi je:nu nugga da:rika da:ri ille.*
> The one we call *nesari*, the small one, that's the eldest brother. Its younger brother is the *hejje:nu*. Younger than him is the *kaḍḍi je:nu*. Younger than him is the *tuḍuve je:nu*. Once, at the time of *gauri* festival, [the eldest] gathered a great deal of food, and said, "Younger brothers, come to my house to eat!" The younger brothers came and had a look—but they couldn't get inside, because the door wasn't right. The *tuḍuve, hejje:nu* and *kaḍḍi je:nu* had no way to get in.

The point of this story is not exactly clear to me, but it might be a way of explaining why the various bees never accidentally or 'deliberately' enter the hive of another species.

[7] Following similar reports of honeybee migration from local communities in many parts of tropical Asia, scientists have confirmed that colonies of *Apis dorsata* do embark on annual migrations of up to 200 km, and faithfully return to the same tree the following season.

7.5.1.2 Thuḍuve je:nu

Thuḍuve je:nu (variously known as the Indian, Asiatic or Eastern honey bee, *Apis cerana*) can be found in both evergreen forests, *ka:nu ka:ḍu*, or in dryer lowland regions, or *na:ḍu ka:ḍu*. With this species as well, people living in a particular locality have a knowledge of the place names associated with the harvesting of *thuḍuve* honey: *togaraṭṭi oṭṭu* and *beḷa:ga:na ka:ḍu*, for instance. Unlike *hejje:nu*, with its preference for particular tree species (or even individual trees), some informants stated that *thuḍuve je:nu* was not picky about its nest site:

> *Adakka mene sikkida:ge, e:va mara gamba i:teyo, oḷa:ge adu ira:de, anta anta gambiga e:va mara uve sariye... ella:ka uve, onde he:ḷakka:ga:dille*
> Whenever it finds a home, any tree hollow (will do), they'll be inside, any tree with such hollows is fine...all of them, I can't name just one.

Thuḍuve je:nu will also readily nest in small rock crevices, even close to ground level. However, when asked to be more specific, some informants replied that this bee can often be found on *ne:ri* (*Syzygium* sp.), *bejja* (*Anogeissus latifolia*) and *karava:di* (*Persea macrantha*) trees, and especially on the *koṭṭa:na beṇḍe* (*Kydia* sp.?).

Two kinds of *thuḍuve je:nu* are recognised by the Solega—*kencu thuḍuve* (the 'red' variety) and *kari thuḍuve* (the 'black' variety). These putative subspecies are said to occupy rather different ecological niches: *kencu thuḍuve* is to be found primarily in the dryer lowland forests, and on *bejja*, *kaggali* (*Acacia chundra*) and *ka:rase* (?) trees, whereas *kari thuḍuve* is said to occur in higher-altitude rainforests, or *male ka:ḍu*, usually on *beṇḍe*, *bejja*, *soravilu* and *puḍu ma:vu* (poss. *Cinnamomum malabatrum*) trees, and in rocks.

Thuḍuve je:nu nests are similar to those of the European honeybee (*Apis mellifera, Apis cerana*) in their architecture, and consist of parallel sheets of comb, separated by narrow gaps. As mentioned above, they are constructed within protective enclosed spaces, such as tree hollows and rock crevices. Brood, pollen and honey can all occur on a single sheet of comb, and my Solega informants spoke of a sequence of events occurring on such sheets over the course of 2 or 3 months. To begin with, a comb sheet contains but a little honey. Eggs are then laid in the empty part of the comb, and brood is reared. As the brood emerges, the bees start collecting large quantities of pollen, and then honey. Towards the end of the sequence, it is common to find comb which contains only honey. This sequence might be repeated three or four times in a year.

7.5.1.3 Kaḍḍi je:nu

Another bee species whose honey is eagerly sought after is *kaḍḍi je:nu* (the dwarf honeybee, *Apis florea*). This is the smallest of the *Apis* species known to the Solega, and can be found in all forest types.

Kaḍḍi je:nu is said to appear around the time that the *maruḷi* plant (*Indigofera* sp.) is in flower. When the *iṇḍã* trees (*Pterolobium hexapetalum*) are in bloom, however, the dwarf honeybee appears in the lowland forests, and in particular in the region where hill slopes meet flat land (*orre ga:ḍu*). Certain conditions need to be fulfilled, however, for this bee to nest in a particular location: according to the Solega, a place needs to be open, i.e. not densely wooded (*bailu*), and cool (*shi:ta*), for *kaḍḍi je:nu* to take up residence there. In locales where such conditions exist, this bee will be found on bushes, on clumps of mistletoe (*uppilu*) growing on *bejja* trees, and even on bamboo canes.

The name *kaḍḍi je:nu* recalls the fact that this species' comb completely encloses part of the small twig, or *kaḍḍi*, to which it is attached. Nests of the dwarf honeybee also contain a clearly demarcated honey-storage section, namely the uppermost section of the comb. The part of the comb below the support branch is used for brood.

7.5.1.4 Nesari je:nu

One species of stingless bee is known to the Solega. *Nesari je:nu* (*Trigona iridipennis*) is a tiny black bee, which is frequently found in tree hollows. There was some disagreement regarding its preferred habitat, with some Solega claiming that it occurred in all forest types, and others stating that it was more common in the dryer, lowland forests, especially at the hill slope-flat land interface, and in less densely wooded areas.

Nesari je:nu can be found on *bejja* and *ka:rasa* trees in dry, open country (*begga:ḍu*), as well as on *asuvara* (*Commiphora caudata*). It is possible to find several (5–10) colonies on a single tree. The *nesari* honey, to which powerful disease-combating properties are ascribed, starts to be available from the time that the *te:ku* (teak; *Tectona grandis*) and *beṇḍe* trees are in flower (usually November–December):

> *Je:nu ma:tra aidu a:ḷa tinna:ku, adu tindale, ella...mansarige e:na ka:yali irtadella, mugiya oṇṭōtuule.*
> You need to eat five *a:la* of pure honey, if you eat that, all…the diseases that afflict a person, they go away.

The nests of this stingless bee are difficult to observe directly, because of their small size, and their location within tree hollows and rock cavities. However, the Solega maintain that the brood area is separate from the honey and pollen storage areas.

Excursus 7.3: Types of Honey

Although Aristotle noted that bees collect honey from plants like thyme, and in fact, "[from] every flower that is furnished with a calyx or cup, and from all other flowers that are sweet-tasted", he was convinced that honey is an atmospheric substance deposited on flowers at certain times of the year, and cited as evidence, a lack of temporal correlation between the blooming of flowers and the production of honey:

> …honey is distilled from dew, and is deposited chiefly at the risings of the constellations or when a rainbow is in the sky: and as a general rule there is no honey before

the rising of the Pleiads. (The bee, then, makes the wax from flowers. The honey, however, it does not make, but merely gathers what is deposited out of the atmosphere; and as a proof of this statement we have the known fact that occasionally bee-keepers find the hives filled with honey within the space of two or three days. Furthermore, in autumn flowers are found, but honey, if it be withdrawn, is not replaced; now, after the withdrawal of the original honey, when no food or very little is in the hives, there would be a fresh stock of honey, if the bees made it from flowers.) (*History of Animals,* V: 22)

The Solega have no doubt that honey (or at least its precursor) is produced by the same flowers on which honeybees forage. Several different types of honey are distinguished, including the now rare *nela hu:vina je:nu* 'honey from groundcover flowers' (see Chap. 5 for further discussion). Other honeys include the highly viscous *matti hu:vina je:nu* '*matti* flower honey', which crystallises readily, *ne:ri hu:vina je:nu* '*ne:ri* flower honey', which is also thick, *honne hu:vina je:nu* and *beṇḍe hu:vina je:nu* which are watery (the latter is greenish), and the *manasi:ge hu:vina je:nu*, which also quickly turns into *je:nu shakkare* 'honey sugar'.

7.5.2 *Aspects of Bee Life History*

The most impressive aspect of the honeybee TEK of the Solega is the detailed and in-depth awareness of the life cycle of honeybees, including, in particular, astonishingly accurate elements of honeybee reproductive biology. Astonishing, because unlike many other animals, looking between the legs of an insect like the honeybee tells an observer remarkably little about its gender, and therefore its potential role in reproduction. Moreover, the sexual habits of the reproductive members of a honeybee colony are extremely hard to observe,[8] and any long-term attempts at observing the inner workings of a hive—either within the dark confines of a crevice or tree hollow, or high up among the branches of tall trees—would be met with considerable indignation by hundreds, if not thousands, of well-armed, well-coordinated, flying insects.

Such difficulties are reflected in the fact that the knowledge of bee genders in Europe—whose cultures have a long history of beekeeping—did not emerge until the late seventeenth century, when Dutch biologist Jan Swammerdam decided to look at the internal organs of the so-called 'king' bee under the newly-invented microscope, and discovered that 'he' had ovaries. Until then, it had been widely

[8] In the European honeybee *Apis mellifera* at least, this occurs once in a queen's lifetime, when she leaves the hive for a 'nuptial flight'. During this time, she is eagerly sought out by drones that detect her pheromones, chase her, and mate with her while in flight; the queen may mate with several drones, and stores their sperm within her body for life.

accepted that only a male could be the leader of a hive, and the honeybee colony was often used in political and sociological writings of the time as an allegory of kingly power, wise and benevolent rule, loyalty, industry and a unity of purpose [219]. Confining himself to a discussion of biological gender, Prete [220] relates how the association between bees and kingship was common in other beekeeping societies such as ancient Egypt, and how, even in sixteenth century Britain,

> …an increasing number of seemingly anomalous discoveries about honey bees forced the authors of beekeeping texts to deny or distort scientific findings in order to continue to use the honey bee as a metaphor for the ideal English society; some authors continued to do so, however, going to fantastic explanatory extremes, until the sheer weight of scientific evidence to the contrary could no longer be denied. (p. 117)

It is against this backdrop of western intellectual tradition that I wish to showcase the honeybee knowledge of the Solega.

7.5.2.1 Honeybee Gender and Reproduction

Individual worker honeybees are called *kunni* in Solega, which is also the word for 'girl', while the 'leader' of the hive is called *ra:ṇi*, or 'queen'. This is consistent with the fact that at any given time, most, if not all, the insects in a honeybee colony, including the queen, are biologically female. Already, it is clear that certain basic facts that eluded the beekeeping societies of Europe are known to the honey-gathering Solega, even in the absence of technological developments such as microscopes and observation hives.[9] A skeptical reader might object to my claims on the grounds that my consultants may have internalised such knowledge following conversations with researchers from the ATREE field station, some of whom have worked on honeybee-related studies, and have employed Solega field guides in the process. I present two lines of evidence to counter this claim: first, the younger of my consultants, who are employed by ATREE deny that this knowledge originated from non-Solega people, and insist that it was taught to them by their parents; and second, even older Solega consultants who live in interior settlements away from Kannada (or other) influences were able to verify the views of the younger Solega, in spite of having had little contact with the outside world.

Throughout much of this book, I have tried to avoid making explicit comparisons between the information provided to me by my consultants and the corpus of 'western' scientific knowledge that I have acquired through years of university education. That is because a proper appreciation of Solega TEK requires that it be understood as a coherent and independent system of knowledge—one that exists not in relation to other knowledge systems, but is instead founded on the very observations that Solega speakers repeatedly make in the everyday lives, and the mental categories

[9] A hive with a clear glass (or recently, Perspex) wall that allows observation of the interior of the colony.

that their language provides. In this chapter, however, I am faced with a paradox: in order to demonstrate how 'accurate' Solega honeybee TEK really is, I am forced to map it onto what is known about honeybee biology to 'western' scientists. It goes without saying that this is an unfair comparison, as the Solega lack the mass of instrumentation and formalized methodologies that scientists take for granted, as well as the societal resources to allow individuals to pursue the detached, professional observation of natural phenomena.

As has been made clear previously, the Solega have the additional disadvantage of not being beekeepers, and being totally dependent on the seasonal migration of honeybees. Their observations, then, are based on the frequent, but brief, chance encounters they have with bees when out foraging, or the longer, but less frequent periods of scrutiny when the honey from bee-trees is systematically harvested. Beekeepers, on the other hand, would have far more opportunities to tend, examine and manipulate several hives, which would be available year-round. What is required to even out the odds in favour of the hunter-gatherers is a comparison with the TEK of a non-industrial culture that indulges in beekeeping. Detailed descriptions of bee TEK are hard to find, but an ideal candidate in this respect is the writing of Aristotle, widely regarded as the 'father of natural history'. In his books, *Generation of Animals* and *History of Animals*, one finds sections where Aristotle presents the honeybee TEK of his Greek contemporaries (some of whom are beekeepers), and analyses this information to deduce certain features of honeybee biology. Of course, Aristotle had the advantage of being able to lead a life of leisure, and of having beekeepers to consult with. Still, it seems reasonable to assume that his observations, and those of his contemporaries, were made with little more than the basic human senses, blended with a healthy dose of deductive reasoning.

But first, the scientific facts: as mentioned above, the queen and all the workers (her daughters) are female, while only the drones are male. The queen mates with one or more drones on nuptial flights outside the hive, and produces workers from fertilized eggs, and drones from unfertilized eggs (Fig. 7.1a). When a queen leaves the hive, or is lost, the hive will descend into anarchy unless new queens, produced by the old one, start to develop. Otherwise, the female workers lose their pheromone-induced physiological inhibitions, and start laying unfertilized drone eggs. The worker population crashes due to a lack of new fertilized eggs, and the colony perishes.

Aristotle takes it as a given that the leader of a hive is a "king" (Fig. 7.1b). He is aware of the existence of two other types of individuals—"bees" (workers) and "drones"—among a colony's members, but on the topic of gender, in Book III:10 of *Generation of Animals* [221], Aristotle has the following to say about certain hypotheses that were being offered by other commentators:

> Nor is it reasonable to hold that "bees" are female and drones male; because Nature does not assign defensive weapons to any female creature; yet while drones are without a sting, all "bees" have one. Nor is the converse view reasonable, that "bees" are male and drones female, because no male creatures make a habit of taking trouble over their young, whereas in fact "bees" do.

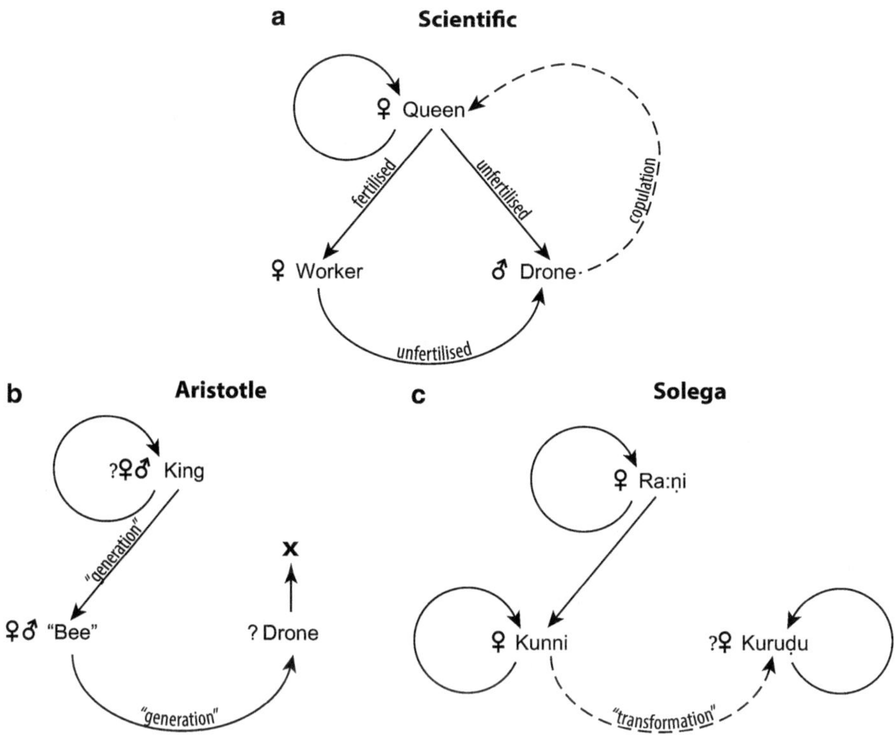

Fig. 7.1 The three conceptions of honeybee reproduction and behaviour described in this chapter. Previously published in [148]

His rejection of the first hypothesis, we now know, was far too hasty. His detailed and accurate descriptions of many aspects of honeybee natural history notwithstanding, Aristotle failed to gain a complete understanding of honeybee reproduction, possibly due to the fact that it is very difficult to observe a drone copulate with a queen. Hence, he writes:

> The generation of bees is a great puzzle…either (i) each kind generates its own kind, or (ii) one of the three kinds generates the others, or (iii) one kind unites with another kind.

He adds that "bees" (and possibly also the king, but the language dealing with this point is vague) contain within themselves "*the male as well as the female, just as plants do*", and that they are able to *generate* offspring without recourse to copulation. Eventually, however, by eliminating various untenable possibilities, and by drawing heavily on key observations of beekeepers, such as the following,

> …the brood of the drones is produced even when there is no drone present to start with, whereas young "bees" are produced only if the kings are present…

Aristotle concluded that:

> …the leaders generate their own kind and another kind as well, (viz. the "bees"); while the "bees" generate another kind (the drones), but *not* their own kind; this they have been deprived of doing. And since any business of Nature's always has an orderly arrangement, on that account necessity requires that the drones shall have been deprived even of generating some other kind. And this is what is found to be the case in actual fact: they are generated themselves, but generate no other creature…

The Solega possess detailed information on the breeding schedules of at least some of the four named bee species. Moreover, there is a clear understanding that the pollen and nectar gathered by the colony are for the purpose of nourishing new brood. This was made clear by explicit statements from informants that the intensive collection of honey and pollen tended to accompany the rearing of brood. In fact, it would be unusual to find honey in a hive which did not also contain some amount of brood.

Unlike the 'scientific' and Aristotelian conceptions of honeybee reproduction, the Solega believe that each honeybee caste is able to generate other individuals like itself (Fig. 7.1c). There are complications, however, because the *ra:ṇi* 'queen bee' is above all the *awwe* 'mother' of all the bees, and is responsible for, presumably, the first generation of *kunni* 'daughter' (worker bees) in a newly established colony. Moreover, the drones can be generated by another mechanism, namely the transformation of *kunni* into the fatter, stingless, unproductive counterparts through the loss of a sting. The following six extracts from three speakers sum up the Solega position on the origin of honeybee castes:

1) *Ra:ṇi noṇa andare, adu ondu ta:yi tara je:nugaḷige…adu jopa:na:gi no:ḍkoḷḷutte, adu ashṭu huḷa iruttella, ashṭu huḷa adu biṭṭu koḍalla, ella huḷa uwe no:ḍkoḷḷutte.*
   As for the queen bee, she's like a mother for all the bees…she looks after them carefully. However many bees there are, she never leaves them, she looks after all of them.
2) *Saṇṇa noṇa ra:ṇiyindave baradu. Adu avara awwa.*
   The small bees [workers] come from the queen. She is their mother.
3) *Moṭṭe iṭṭa:ga, kuruḍu noṇa ade—adu be:re ja:ga ma:ḍutte, alliye, ra:ḍe pakkadalli… ma:muli noṇa ondu ba:gadalli ma:ḍutte. Kuruḍu noṇada tarave, adu e:na moṭṭe iḍitto, ade mari huṭṭidde. Ma:muli noṇada mari ma:muli noṇada mariya:gi huṭṭidde… je:nu gu:ḍu kaṭṭida:ga ra:ṇi adaralli moṭṭe iṭṭurutte. Ra:ṇi moṭṭe iṭṭa:ga a: tara mari a:godde.*
   When eggs are laid, the drone bee—it lays on one spot, there, by the side of the hive. The other bees do it in another spot. Just like the drones, whatever eggs they lay turn into young bees just like themselves. The young of ordinary bees turn into ordinary bees (like their parents)… when a (new) hive is built, the queen lays eggs in it. The queen's eggs hatch into queens just like her.
4) *Uḷukka murtu uḍtade, tikadalliye, tiguṭeli. A: uḷukka murtu uḍtade enda:ga, adaralli innondu uḷukku ille, adave kuruḍu noṇa. Visa kaḷadu uḍtade.*
   The sting breaks off, from its (the worker's) bottom. When the sting breaks off, it does not have another sting—that's how it becomes a drone. It loses its poison.

5) **(K)** *Ulukku tara banda:ga adu kaḷekoḷtade, adu oitu. A: ṭaiminalli kelagondu satto:gutte. Kelagondu iddare adu dappa kunni a:gutte... Su:ji iddare a: tara a:galla... cikka huḷa ja:sti kelsa ma:ḍutte, ya:kandare idu vaisa:gide solpa, idu kaḷekoḷtu, idu solpa kaḍame kelsa ma:ḍutte. A:ga idu mari ma:ḍda:ga doḍḍa ho:ls baratte ra:ḍeyinda. Cikka huḷa ma:ḍdare saṇṇa ho:ls baratte. Idu: be:ga be:ga mari ma:ḍutte, ya:kandare idu kelsa ja:sti ma:ḍutte. Adu nida:na iradarinda ta:katu solpa kaḍame... a: su:ji e:na:daru ho:dare kelsa ma:ḍakke shakti illa. Solpa kaḍame tara kelsa ma:ḍutte. Su:ji iddare ja:sti be:ga be:ga kelsa ma:ḍutte, ya:kandare ella, "ṇavu ondu tara a:ro:gyava:gi iddivĩ" anta.*

   The sting of the (worker) bee is lost, it goes away. At that time, some die. Some that remain turn into fat workers. As long as the needle is there, that doesn't happen... the small bees work more, and because these (drones) are a bit older, and have lost the sting, they work a bit less. And when they lay eggs, there emerge big cells in the hive comb. When the small bees lay eggs, there are small cells. These reproduce quickly, because they can also work fast. The drones are slow, and they are weak... once their sting goes away, they no longer have the strength to work. Their work slows down. While they possess their sting, they work much faster, because they're like, "We're in good health".

6) *Ra:ṇi noṇa ya:va ja:gadalli ku:tiddado, a: ja:gadalli ma:tra ra:ṇi noṇa utupatti a:dde.*

   New queens will emerge from only those spots (cells) that the (existing) queen has sat on.

Extract 1, 2 and 3 indicate that while each caste can generate its own kind, the queen is ultimately the progenitor of the hive. Extracts 4 and 5, on the other hand, illustrate the belief that since drones are fatter and less active than the workers, and also stingless, they must be ex-workers who are transformed after losing their sting (the loss of the sting is almost always observed when a worker bee stings a human). The sting, then, is not only a defensive weapon and a source of the bees' intensely algesic venom, but also the basis of their motivation to forage. While these statements echo elements of the deductive logic of Aristotle, it is in fact the last extract (6) which is the most intriguing, and points to possibly the key observation responsible for the elevated status of the *ra:ṇi* 'queen' in the Solega system—the fact that only queens can produce other queens. As a result, it is the queen who is responsible for honeybee reproduction at the level of the hive as a superorganism, a phenomenon which will be discussed in the next section.

7.5.2.2 Swarming

Swarming is a natural process in honeybee colonies, and occurs mainly in response to overcrowding. The existing queen starts to lay eggs in special queen cells, who then develop into new queens—the first to emerge usually kills the rest, and takes control of the hive. In the meantime, the old queen leaves the hive with about half the workers, and occupies a new nesting site, stopping along the way at various places. When a swarm lands at one of these intermediate sites, scout bees fly off in all directions to locate potential nest sites. They return to the main swarm and perform dances that advertise the locations they have found. When many scouts have agreed on a single nest site, the swarm flies off to the new location, and takes up residence there.

Aristotle says little about this phenomenon,[10] but he does make note of the fact that the hive's leaders may sometimes be killed by other leaders, especially during periods of adverse environmental conditions [222].

> Many of their rulers are also frequently killed, and especially the bad ones, in order that the swarm may be dispersed by the numbers. They are more disposed to kill them when the swarm is not fruitful, and no casts are formed. At such times they destroy the royal cells, if any have been prepared, for they are the leaders of the swarm. (Book IX:26)

He also mentions occurrences of the 'king' leaving the hive in the company of many bees, although he neglects to explain why.

> The king bees never leave the hives, either for food or any other purpose, except with the whole swarm; and they say that, if a swarms wanders to a distance, they will retrace their steps and return until they find the king by his peculiar scent. They say also that, when the king is unable to fly, he is carried by the swarm; and if he perishes, the whole swarm dies with him. (Book IX:27)

One of the first discussions of the causes of swarming behaviour appears to have been written by the English apiarist John Geddes in his 1721 monograph *The English apiary, or, The compleat bee-master* [223]. Here, he blames low food stocks and inclement weather, coupled with overcrowding in the hive, for forcing bees to abandon their old nest.

> ...but if the summer be also moist, the increase of *bees* [*sic.*] will be great; but because of the scarcity of honey, this increase will prove a decrease... for the weather keeping them much in, they can do nothing but breed, and when they go abroad, bring in nothing but bee-bread [pollen] and water, wherewith they feed their young, but can find nothing to lay up in store, so that moist weather gives them two causes of swarming, plenty of *bees*, and penury of honey; and so neither winds, nor clouds, nor rain can stay them. (p. 40)

The Solega have quite explicit knowledge of why a swarm leaves its natal colony, and of the events that occur between departure and arrival at a new nest location. Here, too, population increase in the original colony is held responsible for triggering a swarming episode:

> *Ra:ṇi noṇa e:na ma:ḍutte, be:re mari ja:sti ma:ḍidella, a: kuṭumbake ondu ra:ṇi noṇa ce:nj a:gu uḍte. Ra:ṇi noṇa ce:nj a:da:ga be:re kuṭumbaka ba:ga a:godde. Ra:ṇi noṇa mari ma:ḍda:ga ma:tra. Ra:ṇi mari ma:ḍadiddare, adu onde kuṭumba. A: ra:ṇi noṇa ba:ga a:da:ga be:re mane kaṭṭutte ho:gi. A: ra:ṇi noṇa matte a: kuṭumbadalli uve ade ma:ḍutte. A:ge ba:ga:gi ba:ga:gi ja:sti je:nu gu:ḍu a:dde.*

[10] Some translators have made liberal use of the word 'swarm', but the contexts clearly indicate that Aristotle meant 'hive' in most cases, e.g. "The bee will live for six years, some have lived for seven, and if a swarm lasts nine to ten years, it is considered to have done well." ([222], VII:9)

What does the queen bee do? She has produced lots of offspring, and that family gets a new queen. When there's a new queen, the rest of the family is divided (into two). But only when the queen reproduces (new queens). If not, it remains as one family. When one of the queens gets a part of the family, it goes away and builds a new house. That queen repeats the process in that family as well. Thus, by dividing over and over, you get many bee hives.

(K) *I: tara hecca:du je:nu. Ra:ṇi noṇa eshṭu hecciddo, ashṭu kuṭumba uve heccutte. Kuṭumba heccisa be:ka:dare ra:ḍeyinda huḷagaḷu ma:ḍutte. Ra:ḍeyinda huḷugaḷu ja:sti ma:ḍda:ga ja:sti huḷa heccutte. Ra:ṇi noṇa moṭṭe ikki mari ma:ḍda:ga a:va:ga kuṭumba eraḍu ba:ga a:gu uḍte.* That's how bee(hive)s proliferate. However many queen bees there are, that's how many families you get. When (the queen) wants to grow its family, it makes other bees from the comb. When that happens, you get lots of bees. When the queen lays eggs and produces (queen) offspring, the family divides into two.

The second extract presented above contains more or less the same content as the first, but offers, almost in passing, a valuable insight—that 'reproduction' in bees really should be understood as two parallel phenomena taking place on two time scales. The first is the growth of a colony's population, which continues practically every day, and the second is the division of colonies into daughter colonies through swarming, which only happens a few times in a year.

Knowledge of what happens to a swarm after it leaves its original nest site is arguably the most fascinating and impressive piece of Solega honeybee TEK. First, it is said, the swarm will often land on a tree, which serves as a temporary resting place (Fig. 7.2). Such behaviour is also seen after honey has been harvested from a hive by humans, and the surviving bees are forced to look for a new nest site:

Fig. 7.2 A swarm of *Apis dorsata* bees that has settled temporarily on a *Bauhinia purpurea* tree

Speaker A: *Ondu gu:ḍu, mansaru ho:gi kitta:ga a:vaga ra:ṇi matte a: gu:ḍu noṇagaḷu ondu agala ira bahudu, a: gu:ḍinalliye irutte.*
Speaker B: *Onde-eraḍe jina. Matte je:nu innondu kelsa ma:ḍtade, a: je:nu summane ondu maraka ho:gi kaṭṭidde. Adu togaruguḍukã. A:ga alli kaṭṭu uṭṭu patte ma:ḍtade, se:ra be:kada ja:gava.*
Speaker A: *Horra murri, a:je:nu. Adava iḍiyadille.*
Speaker A: When people go and harvest honey from a hive, the queen and the other bees stay in the same spot for a night or so.
Speaker B: One or two days. Then the bees do another thing, the bees simply go and sit on a tree (without nesting). Those bees are called *togaruguḍukã* ('sap drinkers'). Then they land there and make enquiries regarding possible locations to nest in.
Speaker A: Those bees are homeless. We don't catch them.

The exact way in which bees "make enquiries" was elucidated by one consultant from *Bu:ta:ṇi po:ḍu*, who depicted the process as a conversation between a scout bee and a potential host tree.

DG: *Be:re kaḍeyinda je:nu bandu ondu ka:li marakka kaṭṭade. Hejje:nu endu tiṭṭko:ḷi. Hejje:nu bandu ondu marakka ku:tu takshiṇa a: maradalli taḍeyalla. A: maradalli togaru kuḍitade. Togaruguḍukã entivĩ adanna... E:nappa, adara sa:istra? Adu mi:selu. ... togaruguḍukã allinda bandu i: marakka kaṭṭa be:ka:dale ondu ka:raṇa ide. E:ka togaruguḍukã aitu? Adu muṭṭu tiḷsakoḷtade. Mi:selu a:gu uḍtu. Adu muṭṭu tiḷsakoḷakko:skara be:re maradalli bandu kombinalli kaṭṭruttade. Matte a: maradalli kaṭṭidda:geve innu mattu ondu maranna maneya huḍukitta:du ... Huḍuku uṭṭu ondu aidu a:ru si:ṭu, mane no:ḍadu endale ondu mane oḷḷedu, ondu mane keṭṭadu anteḷi adu ho:gi alli sa:istra ma:ḍa:du. Je:niga buddi ide. A: maravanna ke:ḷa be:ku—"Appa, na:vu ishṭu jana ondu maneli iddivĩ, ni:vu ishṭu janina shakti taḍkoḷtiya? Athava taḍkoḷa:re?" Ha:gente:ḷi maravanna ke:ḷa be:ku. A:ga honne mara he:ḷutte, "Appa, ni:vu sa:vira a:ru jana iddaruve na:nu taḍkoḷtini, nanna kombu kombige bandu se:ri kaṭṭkoṇḍru na:nu taḍkoḷtini." Ha:ge de:rya koḍa be:ku. A: de:rya koḍa takshiṇa a:mele avaru hattu jana bandu uṭṭu avaru ke:ḷu uṭṭu a:mele hattu jana hinde va:pas bandu ivarina, "Ã neḍi appa, alli ho:gõ. Alli namma mane ade, neḍi appa". A:mele alli ho:gu uḍta:re. Ho:ga takshiṇa kuntu ondu eṇṭu diniga ha:ḷe ha:kurte.*
DG: The bees come from elsewhere, and land on an unoccupied tree. Let's take the giant bee. When the giant bee (swarm) comes and sits on a tree, it does not stay there long. It drinks the sap on that tree. That's when we call it *togaruguḍukã.* And as it sits on that tree, it also looks for a home on other trees. What does it do next? It's fussy. There's a reason why the *togaruguḍukã* comes from elsewhere and lands on a particular tree. What causes it to become *togaruguḍukã*? It finds things out by touch. It's become fussy. It goes to another tree and sits on a branch in order to find things out through touch. And then it moves on to another tree again. And when it moves, five or six individuals (first) go to a tree to determine whether it would make a good home or not. The bees are smart. They ask the tree, "O tree! There are many in our household; will you have the strength to be able to bear us all? Or will you not?" That's what they need to ask the tree. Then the *honne* tree says, "O bees! I will support all 1,006 of you; come and sit on all my branches, for I will support you." That's how it reassures them. As soon as they're reassured, those bees go back, and say to the rest, "Yes, come on, let's go! Let's go, our home is there!" They all go there (to the new tree). At the new tree, after eight day or so, they start to build new comb.

Excursus 7.4: Terms for Stages of Comb Development

Once a swarm arrives at a new nest location, the workers start to build the first precursors of the future hive, which will ultimately contain all the food stores (honey and pollen) and brood cells (for rearing larvae), and give the colony a substrate on which to rest. The very first rudiments of comb are a dazzling white, being constructed from newly-secreted wax, and untouched by any impurities. As the comb grows, the appearance of the wax and the overall structure changes. The Solega have terms that describe different stages (and different parts) of the comb, depending on whether it is new or abandoned, or the nature of its contents.

(h)a:le	New comb
bella:ḷe	Newly constructed (white) comb
u:ṭe	Comb with less honey
eralu	Brood comb
tale	Upper part of comb, bearing honey
ra:ḍe	Comb (general), or part of comb without honey
mudi	Colony just arrived at new nest site
iṭṭu kodaṇi je:nu	Comb with lots of pollen
ta:vilu	Comb full of honey? drone comb?

7.5.2.3 Foraging

Other aspects of honeybee lore, not touched upon by Aristotle, are also mentioned by the Solega. Prominent among these is the fact that some bees can forage at night, a phenomenon not reported for the genus *Apis* in the scientific literature until the mid-1980s [224]. The following two extracts illustrate not only this ability of the bees, but also the fact that they often forage over long distances (a few kilometers, in this case).

Kammuri bareyalli je:na:de. Sa:gaḍe mara gummana guḍḍeli ade. Ashṭu du:radalli ade adu. A:ga beḷakina ja:vã ṭaiminalli hejje:nu ella ḍama ḍama ḍama ḍama enduru. A: hejje:nu yelli ade? Kammuri bareyalli ade, ashṭu du:rinda banda:de, kelsa ma:ḍa:ge, kattaleliye.

There are bees on *Kammuri* rock. The *sa:gaḍe* tree (*Schleichera oleosa*) is to be found on Owl Hill. It's that far away! Around the time of sunrise all the *hejje:nu* fly (to the tree), going, "*ḍama ḍama ḍama ḍama!*" Where are those bees (from)? *Kammuri* rock, they come from so far away, just to work, in the dark.

Speaker A: *Je:nu innu ondu kelsa ma:ḍutte. Ba:ninalli moḍa ondu cu:ru ira ba:rdu. tingaḷa biḷuka ira be:ku. Iruḷinalli tingaḷa biḷuka iddare kelsa ma:ḍutte je:nu. Hejje:nu ma:ḍutte, tuḍuve uve ma:ḍutte.*

Speaker B: *A:ga e:va hu: uve ille, sa:ga:ḍe hu: ma:tra*

Speaker A: *Sa:ga:ḍe ta:ri hu:*
Speaker B: *A:ga ra:tri uve je:nu kuiya:du]*
Speaker A: The bees do another thing—it should be a totally cloudless night. There needs to be moonlight. When it's a moonlit night, the bees get to work. The *hejje:nu* does this; *tuḍuve je:nu* also does this.
Speaker B: At that time, just the *sa:gaḍe* flower is in bloom, and no other.
Speaker A: The *sa:gaḍe* and *ta:ri* flowers.
Speaker B: The bees forage then, at night.

Speaker A: *Je:nu eshṭu hu: me:le kelsa ma:ḍiddado, ashṭu hu: senda:gi phala koḍutte. Je:nu ya:va maradalliyo, marada hu:gaḷa me:leyo, athava saṇṇa sette, saṇṇa giḍagaḷa me:leyo, adu ondu ba:ri eraḍu ba:ri hu:vina me:le ku:tidare adu phala ja:sti baratte.*
Speaker B: *I:ga ra:gi tene—ra:gi hu: banda:della, a: hu:ge eshṭu je:nu ku:tatto, ashṭella oḷḷe ka:ḷu a:godde.*
Speaker A: *E:kandale ya:va hu:viga uve adakka gaṇḍa illa ante. Hu:vina me:le ya:va siṭṭe noṇa ho:gi ku:tado, adutte gaṇḍa, a: hu:vige. A:va:ga phala ja:sti... Ya:va hu:vina me:le je:nu illavo, a: hu: ka:yi buḍalla... Siṭṭe gaḷu ira bahudu, dumbi gaḷu ira bahudu. I:ga avare ha:kirtivi na:vu. Jami:ninalli tumba avare ide namage. A:va:ga ondu sari hu: buḍutte, avare hu:. Alli no:ḍadakke ho:dare bari dumbigaḷe ja:sti.*
Speaker B: *A: dumbi avare hu:va kuiya:ge. Kari dumbi adu. Ishṭo dumbi hu:va kuidattu enda:ga varshadalli santoshada avare!... Avare hu:viga dumbiye gaṇḍa. I:ga, obba hengisu. Hengisiga gaṇḍu ho:gi muṭṭida me:le tane phala baradu. A: tara. Heṇṇu gaṇḍu a:da:ga tane magu huṭṭadu. A:ga dumbiyo thuḍuve noṇavo alli ku:ta:ga bemmansi a:goitade.*
Speaker A: *I:ga ju:nu tingaḷalli sakkat honne hu: buḍutte. Honne hu:vigave ondu tarada je:nu kaṭṭutte. Honne hu:vina je:nu adu.*
Speaker B: *Hesare ade honne hu:vina je:nu enduru.*
Speaker A: *Honne hu: me:le a: je:nu bandu kelsa ma:ḍdare a: mara ja:sti honne ka:yi buḍutte.*
Speaker B: *O:! I:ga no:ḍe be:ku i:ge ade honne ka:yi! O:hohoho! Adu enta phala koḍa:du adu... Onde hu:va buḍalla je:nu, je:nu muccidade.*
Speaker A: All the flowers that bees work on go on to produce good crops of fruit. No matter what plant—be it the flowers of trees, or small herbs, or small plants—if bees sit on them once or twice, they yield a lot of fruit.
Speaker B: Take *ra:gi* inflorescences—if even a few bees sit on them, they will give a good harvest of grain.
Speaker A: Because it is said that there are no males among flowers. The butterflies and other bugs that go and sit on flowers themselves become the males for those flowers. That's when you get many fruit. A flower that is not visited by a bee will not bear fruit. It could also be butterflies, or beetles/other insects. We plant beans, for instance. We have many bean plants in the ground. At a particular time, bean flowers blossom on these plants. If you go and take a look then, you'll see lots of large bugs on the flowers.
Speaker B: Those bugs are foraging from the bean flowers. They're black beetles (solitary bees?). If you see a lot of these bugs on the plants, that means a bumper harvest of beans that year! The bugs are the males for the bean flowers. Consider, a woman. She bears fruit only after a man goes and touches her. That's how it is. It's only after they're together that a child is born. So when a beetle or a honeybee goes and sits (on the flower), (the flower) becomes pregnant.
Speaker A: Now, the *honne* flowers bloom in large numbers in June. A particular kind of bee is found on *honne* flowers. That's the *honne*-flower-bee.

Speaker B:	That's its name—the *honne*-flower-bee.
Speaker A:	When this bee works on the *honne* flowers, that tree will produce lots of fruit.
Speaker B:	Oh! You should see the masses of *honne* fruit! Ohohoho! So numerous are the fruit—the bees don't leave a single flower untouched!

The lively account in the above extract is not precisely a description of plant sexuality *per se*—the role of pollen in the fertilization process is not mentioned, and the speakers seem to agree that there is no male counterpart of the fruitful flower. Still, the conclusion that visitations by insects represent the transfer of some kind of male element to the flower (Speaker B makes the unambiguous analogy of the act of copulation between a man and a woman resulting in a child) is an impressive one, given that the Solega regard this principle as applying to all pollinating insects and all flowering plants. In contrast, although Aristotle and his student Theophrastus were aware of a correlation between bees visiting olive trees and heavy crops of olives, the artificial pollination of date palms and the role of fig wasps in the production of edible figs [225], their conception of 'male' and 'female' plants was actually based largely on non-sexual utilitarian factors, and in the vast majority of cases was quite different from modern biological notions of the term [226]. In the case of the date palm, Theophrastus likens the process of pollination to the fertilization of eggs by the milt of a male fish, but adds that this case is exceptional, taking place in two plant species at most.

7.6 Conclusion

The Aristotelian conception of honeybee natural history consisted of accurate insights based on available evidence, as well as (at the time) logical, educated guesswork, in those instances where key evidence was not readily available. The Solega seem to have built up their picture of honeybee biology on identical principles. As a different evidence set was available to them, however (the Solega only gather wild honey from the forest; they have never been beekeepers), their final notion of honeybee natural history must necessarily differ from that presented in Aristotle's writings, who frequently cites apiarists' reports to back up his claims. Nevertheless, their interactions with wild honeybees over countless generations have enabled the Solega to attain an understanding of this important insect's behaviour, migration, reproduction and ecology, which is totally consistent with their observations and experiences. It is truly remarkable that they recognise the queen and foragers as female, that they acknowledge of the role of the queen in swarming and migratory behaviour, and that they are aware of the behavioural and anatomical differences between foragers and drones, all in the absence of a beekeeping tradition. As Ellen [72] has pointed out, the strict academic demarcation between 'hunter-gatherer' and 'agriculturalist' does not accurately portray the activities of groups so labelled, as 'hunter gatherers' often manage and cultivate plants, while 'agriculturalists' frequently make use of gathered natural resources. It makes little sense to predict evolutionary trends of 'complexity' on the basis of small sets of synchronic data, especially when the two groups being compared exist only as theoretical constructs. A proper appreciation of the ethnobiological knowledge of a people can only be gained by investigating not hierarchies of lexemes, but what Ellen calls the people's *depth of knowledge*, which might include knowledge of the basic biology of individual organisms, (utilitarian) knowledge of the natural world that enables humans to gain a fitness advantage, knowledge of general principles learnt through observation, and accumulated over several generations, and finally, systematic encyclopaedic knowledge that connects individual observations of natural phenomena into a single system. Such an investigation, presented in this chapter, has shown that the Solega clearly have a deep understanding of bee natural history, in spite of—as Berlin would have it—their perceptually disadvantaged status as gatherers of honey.

Chapter 8
Conclusions

8.1 The Nature of TEK

The publication of "the ethnoclassification of taxon X in language L" can give a reader the impression that such a classification scheme represents an accepted norm in the speech community, and that an average member of that community would have a mental representation that resembled, at least closely, the abstract scheme presented by the researcher. The issues surrounding Solega bird classification, discussed in Chap. 4, and to a lesser extent, the semantics of the forest terms, discussed in Chap. 5, suggest that caution is required in generalising from the speech or TEK of a handful of individuals to the linguistic norms or TEK of an entire community. In his investigation of variability in Dene bird nomenclature and classification, Gardner [68] reported phenomena that are strikingly similar to those discussed in Chap. 4, such as the idiosyncratic coining of names, considerable phonetic variation in bird names (in spite of the community only having 120 adults), lack of knowledge of the name of certain important species, and lexical variation that went far beyond the merely phonetic. Faced with this level of variation Gardner, in his "requiem for the omniscient informant", concluded tentatively that although members of a community share certain beliefs and a certain body of knowledge, they also exhibit considerable cognitive variation. Reacting to what he saw as too pessimistic a view of the ethnographic enterprise, Boster [67] showcased the knowledge, among the Aguaruna Jivaro, of the over 60 varieties of manioc cultivated for food. Based on his experimental results involving a large number of consultants, Boster concluded that although variation was present at different "layers of social identity", namely 'adult member of the society', 'mature woman', 'close kinswoman', and 'the individual', there was nevertheless a single culturally accepted model of manioc categorisation.

Boster makes a compelling case for the existence of a core body of manioc-related knowledge, with little inter-individual idiosyncratic variation, but unfortunately his study ignores a key variable—the fact that manioc is an intensively

A. Si, *The Traditional Ecological Knowledge of the Solega*, Ethnobiology,
DOI 10.1007/978-3-319-24681-9_8

manipulated food crop, whereas the interactions between the Solega, for instance, and the bird species that live in their forests are far more opportunistic and long-distance. A similar conclusion is reached by Ellen [72], who argues that the classification of frogs (not a food source) in Nuaulu is characterised by lexical variability and uncertainty on the part of his consultants, because the depth of knowledge that people have on any topic is determined by subsistence need, as much as it is by perceptual salience. On the other hand, Berlin has made some strong claims regarding the perceptual status of cultivated crops and the putative effect they have on the appearance of sub-generic taxa in a community's language. I have indicated in Chap. 7 that I am not in favour of such generalisations, but I do believe that it is far more likely that a community will have a 'standardised' body of knowledge (including means of identifying sub-taxa) in the case of a highly manipulated group of organisms such as a food crop, than for undomesticated creatures such as birds. TEK in the latter situation is likely to be far more variable, as Gardner and Ellen have described, and possibly even far more clumped in its distribution. By 'clumped', I mean that not everyone would have the same level of knowledge as everyone else, and that there could be individuals recognised as 'experts' in the community that others would defer to. Such expertise can come about through culturally-sanctioned division of labour, chance events that lead an individual to specialise in a particular domain, or simple interest and aptitude, which allow and individual to concentrate on a topic that s/he enjoys [227]. Similarly, social, cultural and ontogenetic circumstances can lead to a lack of expertise, or even basic knowledge about certain domains.

Solega villages can be physically separated from one another by large (to walk) distances. As mentioned in the Ethnographic Sketch (Chap. 1), people from distant villages might not see one another except at festival times or family celebrations. Such a situation is ideal for the divergence of not only village-based dialects (although I have yet to study this phenomenon systematically), as well as village-based bodies of traditional knowledge. The variation in bird names and classification reported in Chap. 4 provides clear evidence that each village may have its own set of lexemes that denote natural kinds (albeit with considerable overlap with neighbouring villages). Of course, physical separation is not a prerequisite for variation in TEK—Gardner's study, mentioned above, was carried out among a small community living in the same locality, and yet uncovered high levels of inter-individual variation in bird naming and classification schemes. It could be argued that geographic isolation might only exacerbate such tendencies, leading to even higher levels of variability in communities such as the Solega. Two factors external to TEK need to be kept in mind when discussing the sort of variation shown in Chap. 4. First, as the Solega inhabit a wide geographical range (high-altitude evergreen forests to scrub forests on the lowlands), people in different locations are likely to be exposed to different suites of plant and animal species. This would obviously lead to a divergence in TEK over time; indeed, it was noted that people from the lowland village of *A:repa:ḷya* were not able to reliably name any of the birds that were restricted to the evergreen forest. Second, the lifestyle change that has come about in recent years due to *Lantana* and social influences from the rest of the country will have impacted Solega villages to varying degrees. The lowland villages would be the most heavily impacted—the Solega language is quickly disappearing from these communities—followed by the highland communities, and finally the interior villages, where the old ways of living are still largely intact. It is highly likely that the loss of language correlates well with a decline in TEK, and this would lead to an appearance of variability. These two factors might introduce a certain amount of noise into my data, but a clear, unambiguous signal that indicates true baseline variation can be detected in the different bird-name lexemes recorded from different villages. Patterns of binominalisation of bird names with the—*hakki* 'bird' morpheme were also highly variable, and it seems unlikely that either of the two external factors could have caused this.

8.2 Diachronic Ethnobiology

TEK is not a homogeneous body of knowledge; nor is it a static repository of information passed unchanged from generation to generation. Individual observations of new phenomena made by a few gradually become known to many, and modify a community's body of traditional knowledge. Some results of Chap. 4 are relevant here—one could easily imagine a variant name of the Paradise Flycatcher from one village supplanting that of a neighbouring village over time, or of a particular way of classifying sunbirds, as imagined by one section of the Solega community, gaining currency in the wider community. Chapter 5 contains a real example of a landscape term that appears to be undergoing semantic change—the meaning of the word *tho:pu*, in the speech of younger Solega has converged with that of its Kannada cognate (grove/orchard), and now refers to a small clump of trees growing in isolation. In the speech of older Solega, however, *tho:pu* still has the original, related sense of 'trees growing close together', and is used as a general label for the tree-dominated high-altitude forests that the Solega traditionally live in. An unremarkable semantic shift, accompanied by a shift in referent, can therefore lead to a radical change, between generations, in how the forest is perceived.

TEK encoded in language can also preserve the memory of past phenomena for future generations. The discussion on place names in Chap. 3 shows how the names of landscape features, old villages and even individual trees can preserve extremely fine-grained information on the distribution on plant species, both historical and current. In [8] as well as in Chap. 6, I mention how the invasion of the Solega's home forests by the woody weed *Lantana* has led to massive and irreversible changes to both the forest ecosystem, and to the Solega's lifestyle. A significant component of Solega TEK now includes the ways in which native plants and animals have been adversely affected by a loss of habitat and food sources. Foremost in this category is the difficulty that elephants now have in locating sufficient food and water, or in even travelling from one part of the forest to another. These frustrations have made these animals irritable and dangerous, and the Solega now have to exercise additional care while going about their daily activities. Many culturally significant understorey plants are now rare or have disappeared completely, and Solega TEK is now heavy with the memories of the sights and smells that were once taken for granted. Medicinal plants that are still not completely extinct have retreated to lower altitudes where *Lantana* has yet to establish its grip, and the Solega have had to update their knowledge of these newer habitats.

Increased contact with the outside world is also leading to an influx of new concepts and methodologies, which could either be incorporated unchanged into a pre-existing traditional framework, or modified to suit indigenous needs and beliefs. Outside influences include new crop plants, agricultural techniques, medical knowledge, religious influences and naturally, all the vocabulary associated with these and other innovations. I noted at least two instances of parts of the *Lantana* plant being used by Solega nowadays—the crushed leaves are applied to the skin following a beesting to reduce inflammation, and the charcoal of *Lantana* branches, obtained from a camp fire, is used to clean one's teeth in the morning (Table 3.4). In the process of destroying much of Solega TEK, then, *Lantana* is also able to make a small contribution towards increasing it.

8.3 Fact or Fiction: TEK from the Analyst's Point of View

I have argued repeatedly in this book that two things are essential when investigating the potentially complex mental representations of natural phenomena. The first is the elicitation of data in contexts that are as natural as possible, in order to ensure that the categories and the relationships between categories so obtained are native to the language under investigation, and not artefacts of the data collection method. Failing this, any conclusions drawn from data obtained through artificial contexts (i.e. interviews, goal-oriented experimental tasks) should reflect the shortcomings of the data collection method, and the biases that may be present in such data should be acknowledged explicitly. In the words of William Labov [73], founder of the field of variationist sociolinguistics:

> Either our theories are about the language that ordinary people use on the street, arguing with friends, or at home blaming their children, or they are about very little indeed. (p. 109)

The second point follows directly from the first: naturalistic speech data should be given priority over elicited data, as it is only through studying the patterns of naturalistic speech that we can gauge the accuracy of our analytical models of people's mental representations. The study of such data also reduce the chances of the analyst applying his/her own ontological prejudices to the analysis of native categories in another language.

The study of categorisation in an ethnographic context is one field of enquiry that is especially prone to context effects and over-analysis, as discussed in Sect. 1.3. The existence of 'black' and 'white' as basic colour categories is often cited as a cross-linguistic universal, but as Wierzbicka [109] has pointed out, not even 'colour' can be assumed to be a concept shared by all languages. Careful examination of the contexts in which certain 'colour' terms are used can reveal that perceptual features of objects other than hue might be the referents of such terms. Studies of folk taxonomy can also benefit greatly from the inclusion of naturalistic speech data, as illustrated by McKnight's [77] example of a superordinate taxon being more salient than a 'generic' level taxon in the context of a Lardil speaker announcing his intention to go hunting. Such culturally relevant, everyday contexts clearly have the potential to motivate utterances that are qualitatively very different from those elicited through interviews. A worrying example of analytical exuberance based on very limited empirical grounds is the 'taxonomy' of ancestral beings in the Australian language Dalabon, constructed by Maddock [228], as a means of explaining the 'anomalous' classification of the emu. Maddock's taxonomy contains two key nodes which, in Dalabon, have no name; there are also no additional data present in the paper to support the four-tiered structure of the taxonomy, and it appears that the entire schema was more an attempt "*to account as economically as possible for the ordering of species*" (p. 102) than an accurate description of Dalabon speakers' mental representation of the categorisation of ancestral beings. Wierzbicka [229] sums up the problem of analytical overenthusiasm with characteristic precision:

> To assume that people in all cultures have the concept of 'sadness' even if they have no word for it is like assuming that people in all cultures have a concept of 'marmalade' and moreover, that this concept is somehow more relevant to them than the concept of 'plum jam', even if they happen to have a word for the latter but not the former. (p. 9)

The concept of a cognitive map as a mental representation of spatial relations—one that is either analogous to, or even like, a cartographic map—can also be challenged when one considers empirical data consisting of people's generalisations about the information that is supposedly encoded in such maps. Baird et al. [230] have demonstrated mathematically that certain methodologies used to elicit cognitive maps significantly 'transform' subjects' mental representations in specific ways; as a result, the authors conclude that "*cognitive science must take seriously the realization that method and substantive theory are inseparable*" (p. 213). Additionally, several empirical studies, using techniques such as distance and direction judgements, map recognition and map reconstruction, have shown that people appear to entirely reorganize spatial information from the external environment when producing mental representations (reviewed in [231]). Tversky proposes a 'constructionist' view of cognitive maps, a view where "*representations of the visual world are constructed, and…systematic errors may be introduced in the construction of representations as well as in retrieval of information from them*" (p. 135). She summarises her stance on cognitive maps as follows:

> As we navigate an environment, or make inferences from memory of one, we draw on information from many different sources, from particular episodes in the environment and schematic knowledge of the environment, from verbal descriptions and visual experience, from information specific to the environment, and from general information about that kind of environment… When all that information is put together, it does not necessarily form a coherent picture, something that could be drawn on paper or modelled in three dimensions. On the contrary, the different bits and pieces may very well conflict with each other, something that would not be evident without an attempt to put them together. (p. 137)

The objections raised by Baird et al. and Tversky against a neat, simplistic and unified view of cognitive maps echo my own doubts about the status of elaborate folk classifications (one for 'Animals' and one for 'Plants') as real, static mental representations shared by members of a speech community. As I have argued in Chaps. 1–4, and as pointed out by other authors [13, 14, 71, 151], folk classifications are inseparable from the social contexts in which they are routinely used, and any effort to elicit a complex taxonomy in an interview or experimental situation will result in the documentation of categories and relationships that are alien to the speakers' usual way of thinking. Far too many studies on classification systems concern themselves with only the lexemes that label biological categories, and the hierarchical relationships between them. Covert categories of natural kinds are frequently posited in classification systems, but these rely heavily on perceptual sorting and grouping tasks, with linguistic data being rarely used. The results of the Boster and Johnson [56] experiment mentioned in Chap. 2 suggest that people will use purely perceptual cues as a last resort when no other information—linguistic or encyclopaedic—is available to them. The case for 'perception' in the construction of ethnobiological classifications is probably highly overstated, and as some authors have argued, the classifications derived from perception-based sorting and grouping tasks are probably not accurate representations of native conceptions of the relationships between natural kinds. Ellen [232] notes that card-sorting tasks suffer from serious methodological flaws (even though the task itself may be a useful way to elicit novel information from respondents), and that any hierarchies generated by this method should be regarded with caution because:

> the human mind is sufficiently flexible to 'rewrite' classifications in a taxonomic form, to rely entirely on binary discriminations, or upon morphological criteria, depending on implicit prompting. (p. 28)

Additionally, the fact that these tasks are carried out by individuals in isolation (i.e. without consultation with other speakers) exposes an assumption that there is a single accepted classification evenly distributed across the speech community, and that uncovering that classification is simply a matter of averaging the responses of an adequate number of members of that community. The Solega bird naming study presented in Chap. 4 refutes such an idea, showing instead that variation is indeed the norm even at the level of lexical labels for common bird species. This variation was uncovered in spite of the fact that in my elicitation sessions, consultants were presented with visual and auditory stimuli as a group, and were allowed to freely consult with one another to reach a consensus.

Similarly, it has to be remembered that cognitive maps of the type described in Chap. 5 (especially Fig. 5.5) are probably not permanently stored mental representations analogous to a cartographic map—it is most likely not a three-dimensional snapshot of the external environment, with herds of elephants and swarms of bees moving across the surface in concert with the seasonal cycle, and accompanied by bursts of flowering first in one habitat, and then in another. Instead, elements of the cognitive map, such as the phenological signature of a particular forest/landscape type, are likely to be reconstructed from the diverse pieces of encyclopaedic information that each Solega speaker possesses, and with a clear goal, such as travelling to a particular location for a specific purpose, in mind.

8.4 Culture and Classifications

An ethnoclassification scheme is merely the skeleton onto which the flesh of a community's TEK is draped. As such, we should be wary of giving too much importance to this abstract construct, given that the true richness of a people's knowledge of the natural world lies elsewhere—in the way they talk about the relationships between groups of living entities, between living entities and the physical world, and between living entities and humans. If an ethnoclassification scheme is to be elicited and analysed, it should be borne in mind that the scheme needs to be considered in its social context, for there are few linguistic phenomena that remain unchanged either in different contexts, or in a decontextualised form. Roy Ellen, in particular, has long been suspicious of folk classifications presented as static crystallisations of people's conceptions of the relations between natural kinds, and has urged fieldworkers to pay greater attention to the amorphous mass of 'culture' that is often rejected in studies of classification (see [233] for a collection of essays that attempt to present "Classifications in their Social Context"). Ralph Bulmer, too, felt that Berlin's predictions (I would prefer to use the label 'prescriptions') were far too restrictive to account for the cross-linguistic variation commonly seen in classification systems [151]—variation that could uncontroversially be attributed to cultural differences. However, when the search for universals turns into a prescriptivist framework into which all data are to be squeezed, variation and anomalies are often ignored, and the resulting classifications start to resemble one another closely.

Berlin is right to say that a Kingdom level taxon can be present even in languages that do not have lexified labels for 'Plant' and 'Animal'. After all, languages such as Tzeltal ([9], p. 193) have classifier morphemes that indicate that this is a very real cognitive division for speakers of those languages. Even in the absence of such morphosyntactic elements that pervade a language, however, Berlin insists that the Plant/Animal Kingdom level taxa are cross-linguistic universals. This is where I disagree, because the fact that respondents *can* categorise all natural kinds known to them into groups that approximate *Animalia* and *Plantae* cannot be taken as evidence that this is how they *usually* perceive the natural world. Moreover, some languages have explicit category labels that speakers will volunteer as indicating the highest-level taxonomic division in their language—a good example is the Australian language Kayardild which distinguishes, with primary lexemes, 'big game', which includes marine mammals and turtles, 'game', which includes non-marine creatures, 'bony fish' and 'sharks and rays' [234]. Surely in the face of concrete linguistic evidence to the contrary, as well as native speaker insistence that "*this is how we divide living things*" it makes no sense to keep arguing that there must necessarily be covert Plant/Animal categories in Kayardild (recall Wierzbicka's earlier comment on 'sadness' and 'marmalade'). If, as Hallpike [190] suggests, categories such as 'things of the forest' and 'things of the sea' are explicitly named in a language (p. 203), is it valid to disregard such groupings in favour of 'plant' and 'animal', which may not be named?

Moreover, if covert categories of organisms that *can* be formed on the basis of morphological similarities are a cross-linguistic universal, then why restrict the scope of this universal to the top of the hierarchy? Why shouldn't languages which lack specific level taxa be re-analysed as possessing generic taxa containing endless numbers of covert specifics? Here I refer primarily to the languages of hunter-gatherers, who, according to Berlin, are unable to notice the finer distinctions between subgeneric groupings (Chap. 7) that agriculturalists have grown accustomed to. I am also reminded of the fact that Solega has no subgeneric lexemes for the large numbers of fern and orchid species that grow in their forests, species that appear distinctive even to casual observers with no biological expertise. The Solega certainly know that there are many kinds of ferns and orchids—are we to gloss these as covert species, because the results of a sorting task would unequivocally yield groupings based on biological species? My answer would be an emphatic 'No!' because the lack of subgeneric labels in this case is the result of a lack of a utilitarian relationship between the Solega and these taxa. The Solega have no use for orchids or ferns, and have therefore not bothered to give them any distinctive names.

8.5 Linguistic Context and Encyclopaedic Knowledge

The notion of 'context' has been shown to play a crucial role in nomenclature and taxonomy (Chap. 4), as well as in the use of landscape terms in naturalistic speech (Chap. 5). In the naming of birds, Solega speakers frequently provided unanalysable mononomial lexemes in elicitation sessions, but subsequent utterances reveal that many such names could in fact be paired with—*hakki* 'bird' to produce a more complete, binomial label, that may be more representative of the way people refer to these birds in naturalistic speech. Further complicating the picture is the fact that (a) birds can have optional binomial, obligate binomial and obligate mononomial labels depending on the species, (b) people from different villages can have markedly different ideas about which birds belong to which nomenclatural category, and (c) other groups of living organisms (when grouped along the lines of Berlin's 'life forms') have very different naming patterns from that of the birds. Clearly, naming conventions can vary not only among the separate sub-populations of the Solega speech community, but also with the nature of the referent being named. Both these findings strongly suggest that the nomenclature of living organisms in Solega is a highly variable exercise, and that in the face of such data, it makes little sense to talk about a unified Solega ethno-taxonomy of birds (or any other taxon), let alone the existence of cross-linguistic universals of classification and nomenclature that are largely unaffected by culture or context.

The Solega landscape terms pattern differently with context—surprisingly, naturalistic speech contains very few references to any forest/landscape type other than the very general terms *ka:ḍu* and *na:ḍu*. These terms serve the very important purpose of contrasting between human habitation and the wilderness in the case of the former, and between Solega and non-Solega territories in the case of the latter. The other terms, however, which encode very fine-grained distinctions between forest habitats, were only used in the context of describing the places where certain organisms or resources are found, or in the context of travelling to (or actively avoiding) certain locations. These contexts suggest that, as proposed by Burenhult and Levinson [104], the interactional properties of the forest/landscape terms (i.e. the nature of the activities that people regularly perform in these places) play an important role in maintaining the distinctiveness of these places for the Solega.

In Chap. 6, I presented an overview of the Solega's rich encyclopaedic knowledge of the ecological processes and relationships that link the plant and animal inhabitants of the B. R. Hills to each other on the one hand, and to the abiotic environment on the other. The annual monsoon cycle provides an oscillating backdrop against which the cyclic behaviours and life histories of living organisms are viewed, while predictable temporal co-incidences between two discrete events, such as the blooming of a flower at a particular time of day, or the migration of a particular honeybee species into the high-altitude forests, inform Solega observers of the passage of time. Certain phenomena will collocate in a single place—certain edible mushrooms are only said to grow on particular substrates—and the Solega use such information to locate useful resources in an otherwise complex forest environment. The Solega are aware of the numerous trophic, agonistic and antagonistic interactions that occur on a daily basis between living organisms, and although such interactions may have a direct bearing on the well-being of humans (e.g. *Lantana* outcompetes native grasses which are the food of elephants; hungry elephants are more likely to attack people), a significant proportion of the interactions noted by the Solega are 'third-party' interactions, where humans play little, if any, role.

The findings in Chap. 6 are complemented by the Solega's keen observation of the utilisation of numerous plant species, presented in Chap. 3 (Table 3.4). Indeed, the number of plants said to be of nutritional or medicinal use to birds and animals (42 species) is close to the numbers of plants used by humans as food (52 species) or medicine (49 species). This interest in the habits of other species arises out of not only utilitarian motivitations, but also the affinity that Solega people feel with other organisms. It is almost a cliché to say that a non-industrial people's way of life is in some way 'close to nature',[1] but in Sect. 6.4, I have presented direct quotes from Solega people professing a strong sense of empathy for all manner of wild creatures ranging from honeybees to elephants. Many Solega folk tales also tell of how certain organisms were once people, and how their current form reflects a relevant aspect of their human personality. Solega encyclopaedic knowledge, then, is the result of not merely a recognition of perceptually salient organisms and phenomena, or an awareness of the utility of particular species and environments, but also a strong sense of empathy with living beings that are perceived to share the same experiences and challenges as the Solega themselves.

Chapter 6 also highlights an ecological approach to linguistic ethnobiology, in contrast to more traditional, taxonomy-focused studies. Hunn ([235], p. 13) has

[1] A Google search for [("native people" OR "indigenous people") AND "close to nature"] resulted in 861,000 hits (July 13, 2015).

pointed out, and rightly so, that "*TEK is* ecological *knowledge*", while Lewis [236], elaborating further on this issue, has stated that the study of TEK

> ultimately proceeds to considerations of processes (functional relationships): the understandings that people have of environmental systems and the networks of cause and effect therein. A part of these understandings involves a people's perceptions of their own roles within environmental systems: how they affect, and how they are affected by, natural processes. (p. 9)

Given the Solega's empathic outlook, it is quite unsurprising that they are so intimately familiar with the hidden world of the honeybee colony, whose inner workings—from the Solega perspective—are described in Chap. 7. Here too, anthropocentric references abound, with the queen bee or *ra:ṇi* 'queen' said to be like the *awwe* 'mother' of all the worker bees or *kunni* 'girl'. The drones or *kuruḍu noṇa* 'blind flies' are lazy and lacking in strength, and many other aspects of honeybee behaviour are explained in terms of human-like responses to the needs and desires that an individual bee or the entire colony might be faced with. The main argument presented in the chapter, however, is that in spite of not being beekeepers, the Solega's knowledge of honeybee behaviour and reproductive biology rivals that of pre-industrial beekeeping cultures. The data presented in this chapter argue against two crucial points made by Berlin: first, that domestication leads to the recognition and naming of 'folk specific' taxa, and second, that hunter-gatherers tend to be less 'careful', less 'systematic' than agriculturalists in the way they look at nature, or deal with the natural world. The Solega clearly have 'folk specific' labels for the four bee species known to them, and the depth of their bee lore rivals, and is in some cases more accurate than, that of Aristotle, who had access to the knowledge of Greek beekeepers. Subsequently, the Solega recognise the queen bee and the workers as females, and know of the twin strategies of honeybee reproduction (producing new offspring and swarming to create new colonies), the behaviour of a swarm and its purpose, the connection between visits by honeybees and fruit-set on flowering plants, and the foraging activity of worker bees as the mechanism by which the 'honey' obtained from flowers is transported to the hive.

8.6 Concluding Remarks

This book began with a discussion of a very mainstream pre-occupation in linguistic ethnobiology—the elucidation of the ethno-classification systems of non-industrial cultures, and the testing of hypotheses concerning the cross-linguistic universality of certain ways of naming and categorising living organisms. I have not only presented numerous counter-examples from Solega to refute some key 'universals' but also questioned, as have other authors, the appropriateness of investigating ethno-classifications outside normal social contexts. Using current developments in evolutionary biology, as well as ongoing debates in biological systematics and taxonomy, I have also argued that some very fundamental assumptions, on which the idea of 'universals' is based, are seriously flawed. The discussion on Solega bird classification not only incorporated contextualiding information, such as geographic location, cultural/encyclopaedic knowledge and utterance type, but also acknowledged linguistic variation, which is an undeniable feature of any language community.

Moving further from ethno-classifications, and deeper into the domain of context and encyclopaedic knowledge, I first gave an account of how Solega people view their (living) physical environment—as a complex, three-dimensional patchwork of habitat types that is constantly being updated along the fourth dimension of time. This mental map serves as the substrate for the countless temporal cycles, spatial collocations, ecological interactions that constitute Solega traditional ecological knowledge, knowledge that provides the Solega with the means to sense the passage of time, avoid danger, find useful resources, and have a general awareness of the state of their forest home. Their brief but frequent interactions with honeybees—a highly significant group of organisms for cultural reasons—has allowed them to reconstruct the complex life history of these insects with remarkable accuracy, in spite of not being able to observe certain obscure but crucial aspects of honeybee anatomy and behaviour. The latter chapters of this book illustrated the point that ethno-classifications represent but a tiny portion of the TEK of a language community, and that ecological, cultural (including religious and utilitarian) and linguistic knowledge can only be separated from each other by radically altering indigenous world views and conceptions. Instead, a language-oriented, context-sensitive approach to the study of TEK can contribute far more to our understanding of such highly integrated knowledge systems.

References

1. Madegowda C, Rao CU. The sickle cell anemia health problems: traditional and modern treatment practices among the Soliga tribes at B.R.Hills, South India. Antrocom Online J Anthropol. 2013;9(2):243–51.
2. Ford R. Introduction - ethnobiology at a crossroads. In: Ford R, editor. Ethnobiology at the millenium: past promise and future prospects. Ann Arbor: Museum of Anthropology Publications, University of Michigan; 2001.
3. Ramstad K, Nelson NJ, Paine G, Beech D, Paul A, Paul P, et al. Species and cultural conservation in New Zealand: Maori traditional ecological knowledge of Tuatara. Conserv Biol. 2007;21(2):455–64.
4. Ban N, Picard C, Vincent A. Comparing and integrating community-based and science-based approaches to prioritizing marine areas for protection. Conserv Biol. 2009;23(4):899–910.
5. Becker D, Ghimire K. Synergy between traditional ecological knowledge and conservation science supports forest preservation in Ecuador. Conserv Ecol. 2003;8(1):1–11.
6. Klubnikin K, Annett C, Cherkasova M, Shishin M, Fotieva I. The sacred and the scientific: traditional ecological knowledge in Siberian river conservation. Ecol Appl. 2000;10(5): 1296–306.
7. Drew J, Henne A. Conservation biology and traditional ecological knowledge: integrating academic disciplines for better conservation practice. Ecol Soc. 2006;11(2):34–42.
8. Si A. Biology in language documentation. Lang Doc Conserv. 2011;5:169–86.
9. Berlin B. Ethnobiological classification. Princeton, NJ: Princeton University Press; 1992.
10. Berlin B. Speculations on the growth of ethnobotanical nomenclature. Lang Soc. 1972;1(1):51–86.
11. Berlin B. Folk systematics in relation to biological classification and nomenclature. Annu Rev Ecol Syst. 1973;4:259–71.
12. Berlin B, Breedlove D, Raven P. General principles of classification and nomenclature in folk biology. Am Anthropol. 1973;75:214–42.
13. Baker B. Ethnobiological classification and the environment in Northern Australia. In: Schalley AC, Khlentzos D, editors. Mental states. Philadelphia, PA: John Benjamins; 2007.
14. Dwyer P. Ethnoclassification, ethnobiology and the imagination. J de la Société des Océanistes. 2005;120–121(1/2):12–25.
15. Bulmer R. Why is the cassowary not a bird? Man. 1967;2:5–25.
16. Bulmer R. Memoirs of a small game hunter: on the track of unknown animal categories in New Guinea. J d'Agriculture Tropicale et de Botanique Appliquée. 1974;21(4–6): 79–99.

A. Si, *The Traditional Ecological Knowledge of the Solega*, Ethnobiology,
DOI 10.1007/978-3-319-24681-9

17. Ellen R. Processing Metroxylon sagu Rottboell (Arecaceae) as a technological complex: a case study from south central Seram, Indonesia. Econ Bot. 2004;58(4):601–25.
18. Gadgil M, Berkes F, Folke C. Indigenous knowledge for biodiversity conservation. Ambio. 1993;22(2-3):151–6.
19. Silvano RAM, Begossi A. Local knowledge on a cosmopolitan fish: ethnoecology of Pomatomus saltatrix (Pomatomidae) in Brazil and Australia. Fish Res. 2005;71(1):43–59.
20. Nombo P, Leach J. Reite plants: an ethnobotanical study in Tok Pisin and English. Canberra: Australian National University Press; 2010.
21. Martin G. Ethnobotany: a methods manual. London: Chapman & Hall; 1995.
22. Berkes F. Sacred ecology: traditional ecological knowledge and resource management. Philadelphia, PA: Taylor and Francis; 1999.
23. Maffi L, editor. On biocultural diversity: linking language, knowledge and the environment. Washington, DC: Smithsonian Institution Press; 2001.
24. Anderson E, Pearsall D, Hunn E, Turner N, editors. Ethnobiology. Hoboken, NJ: Wiley; 2011.
25. Evans N. Sign metonymies and the problem of flora-fauna polysemy in Australian linguistics. In: Tyron D, Walsh M, editors. Boundary Rider. Essays in honour of Geoffrey O'Grady. Canberra: Pacific Linguistics; 1997.
26. Waddy J. Classification of plants and animals from a Groote Eylandt Aboriginal point of view. Darwin: Australian National University North Australia Research Unit; 1988.
27. Telfer W, Garde M. Indigenous knowledge of rock kangaroo ecology in Western Arnhem Land, Australia. Hum Ecol. 2006;34(3):379–406.
28. Himmelmann N. Documentary and descriptive linguistics. Linguistics. 1998;36(1):161–95.
29. Crystal D. Language death. Cambridge: Cambridge University Press; 2000.
30. Nettle D, Romaine S. Vanishing voices: the extinction of the world's languages. New York: Oxford University Press; 2000.
31. Grenoble L, Whaley L, editors. Endangered languages: language loss and community response. Cambridge: Cambridge University Press; 1998.
32. Gippert J, Himmelmann N, Mosel U, editors. Essentials of language documentation. Berlin: Mouton de Gruyter; 2006.
33. Thieberger N, editor. The Oxford handbook of linguistic fieldwork. Oxford: Oxford University Press; 2011.
34. McClatchey W. Ethnobiology - basic methods for documenting biological knowledge represented in languages. In: Thieberger N, editor. The Oxford handbook of linguistic fieldwork. Oxford: Oxford University Press; 2011.
35. Berlin B. How a folkbotanical system can be both natural and comprehensive: one Maya Indian's view of the plant world. In: Medin D, Atran S, editors. Folk biology. Cambridge: MIT Press; 1999.
36. Evans N. In: Crystal D, editor. Dying words: endangered languages and what they have to tell us. Chichester: Wiley Blackwell; 2010.
37. Garde M, Nadjamerrek BL, Kolkkiwarra M, Kalarriya J, Djandjomerr J, Birriyabirriya B, et al. The language of fire: seasonality, resources and landscape burning on the Arnhem Land plateau. In: Russell-Smith J, Whitehead P, Cooke P, editors. Culture, ecology and economy of fire management in North Australian Savannas. Canberra: CSIRO Publishing; 2009. p. 85–164.
38. Majnep IS, Bulmer R. Animals the ancestors hunted. Adelaide: Crawford House; 2007.
39. Majnep IS, Bulmer R. Birds of my Kalam Country. Auckland: Auckland University Press; 1977.
40. Dumbacher J, Beehier B, Spande T, Garraffo M, Daly J. Homobatrachotoxin in the genus Pitohui: chemical defense in birds? Science. 1992;258(5083):799–801.
41. Nash D. Comparative flora terminology of the central Northern Territory. In: McConvell P, Evans N, editors. Archaeology and linguistics: aboriginal Australia in global perspective. Melbourne: Oxford University Press; 1997.

42. Ross M, Pawley A, Osmond M, editors. The lexicon of Proto Oceanic: 3. Plants. Canberra: Pacific Linguistics; 2008.
43. Ross M, Pawley A, Osmond M, editors. The lexicon of Proto Oceanic: 4. Animals. Canberra: Pacific Linguistics; 2011.
44. Wierzbicka A. "Apples" are not a "kind of fruit". Am Ethnol. 1984;11(2):313–28.
45. Pawley A, Syder F. Two puzzles for linguistic theory: native like selection and native like fluency. In: Richards J, Schmidt R, editors. Language and communication. London: Longman; 1983.
46. Boroditsky L, Fuhrman O, McCormick K. Do English and Mandarin speakers think about time differently? Cognition. 2011;118(1):123–9.
47. Winawer J, Witthoft N, Frank M, Wu L, Wade A, Boroditsky L. Russian blues reveal effects of language on color discrimination. Proc Natl Acad Sci USA. 2007;104(19):7780–5.
48. Danziger S, Ward R. Language changes implicit associations between ethnic groups and evaluation in bilinguals. Psychol Sci. 2010;21(6):799–800.
49. Cashdan E. A sensitive period for learning about food. Hum Nat. 1994;5(3):279–91.
50. Barsalou L. Flexibility, structure and linguistic vagary in concepts: manifestations of a compositional system of perceptual symbols. In: Collins A, Gathercole S, Conway M, Morris P, editors. Theories of memory. Hove: Lawrence Erlbaum; 1993.
51. Barsalou L. Ad hoc categories. Mem Cognit. 1983;11(3):211–27.
52. Barsalou L, Wilson C, Hasenkamp W. On the vices of nominalization and the virtues of contextualizing. In: Mesquita B, Barrett L, Smith E, editors. The mind in context. New York: Guilford Press; 2010. p. 334–60.
53. Barrett L, Mesquita B, Smith E. The context principle. In: Mesquita B, Barrett L, Smith E, editors. The mind in context. New York: Guilford Press; 2010.
54. Yeh W, Barsalou L. The situated nature of concepts. Am J Psychol. 2006;119(3):349–84.
55. Medin D, Lynch E, Coley J, Atran S. Categorization and reasoning among tree experts: do all roads lead to Rome? Cognit Psychol. 1997;32:49–96.
56. Boster J, Johnson J. Form or function: a comparison of expert and novice judgements of similarity among fish. Am Anthropol. 1989;91(4):866–89.
57. Ameel E, Storms G, Malt B, Sloman S. How bilinguals solve the naming problem. J Mem Lang. 2005;53:60–80.
58. Malt B, Sloman S, Gennari S, Shi M, Wang Y. Knowing versus naming: similarity and the linguistic categorization of artifacts. J Mem Lang. 1999;40:230–62.
59. Ellen R. Variable constructs in Nuaulu zoological classification. Social Sci Inf. 1975;14:201–28.
60. Sillitoe P. Contested knowledge, contingent classification: animals in the Highlands of Papua New Guinea. Am Anthropol. 2002;104(4):1162–71.
61. Ross B, Murphy G. Food for thought: cross-classification and category organization in a complex real-world domain. Cognit Psychol. 1999;38:495–553.
62. Hunn E. The utilitarian factor in folk biological classification. Am Anthropol. 1982;4:830–47.
63. Forth G. Nage birds: classification and symbolism among an Eastern Indonesian people. New York: Routledge; 2004.
64. Kitchin R. Cognitive maps: what are they and why study them? J Environ Psychol. 1994;14:1–19.
65. Milroy J, Milroy L. Linguistic change, social network and speaker innovation. J Ling. 1985;21:339–84.
66. Labov W. Sociolinguistic patterns. Oxford: Basil Blackwell; 1972.
67. Boster J. "Requiem for the omniscient informant": there's life in the old girl yet. In: Dougherty J, editor. Directions in cognitive anthropology. Champaign, IL: University of Illinois Press; 1985. p. 177–97.
68. Gardner P. Birds, words, and a requiem for the omniscient informant. Am Ethnol. 1976;3(3):446–68.

69. Haiman J. Dictionaries and encyclopedias. Lingua. 1980;50(4):329–57.
70. Hilferty J. Cognitive linguistics: a introductory sketch. In: Lorente M, Alturo N, Boix E, Payrató M-RLL, editors. La gramàtica i la semàntica en l'estudi de la variació (Actes del 5è i 6è colloquis lingüístics de la Universitat de Barcelona (CLUB-5, CLUB-6)). Barcelona: Promociones y Publicaciones Universitarias; 2001.
71. Ellen R. Introductory essay. In: Ellen R, Reason D, editors. Classifications in their social context. London: Academic Press; 1979.
72. Ellen R. Modes of subsistence and ethnobiological knowledge: between extraction and cultivation in Southeast Asia. In: Medin D, Atran S, editors. Folk biology. Cambridge, MA: MIT Press; 1999.
73. Labov W. Some principles of linguistic methodology. Lang Soc. 1972;1(1):97–120.
74. Evans N. Context, culture, and structuration in the languages of Australia. Annu Rev Anthropol. 2003;32:13–40.
75. Simpson J. From common ground to syntactic construction: associated path in Warlpiri. In: Enfield N, editor. Ethnosyntax. Oxford: Oxford University Press; 2002. p. 287–307.
76. Linell P, editor. Towards a dialogic linguistics. XII International Bakhtin Conference; 2005 July 18–22; Jyväskylä, Finland: Department of Languages, University of Jyväskylä.
77. McKnight D. People, countries and the rainbow serpent: systems of classification among the Lardil of Mornington Island. New York: Oxford University Press; 1999.
78. Grice P. Logic and conversation. In: Cole P, Morgan J, editors. Syntax and semantics, 3: Speech acts. New York: Academic Press; 1975. p. 22–40.
79. Myers N, Mittermeier R, Mittermeier C, Fonseca G, Kent J. Biodiversity hotspots for conservation priorities. Nature. 2000;403:853–8.
80. Krishnaswamy J, Kiran M, Ganeshaiah K. Tree model based eco-climactic vegetation classification and fuzzy mapping in diverse tropical deciduous ecosystems using multi-season MDVI. Int J Remote Sens. 2004;25:1185–205.
81. Srinivasa T, Karthikeyan S, Prasad J. Faunal survey of Biligiri Rangaswamy Temple Wildlife Sanctuary. Bangalore: Merlin Nature Club; 1997.
82. Aravind N, Rao D, Madhusudan P. Additions to the birds of Biligiri Rangaswamy Temple Wildlife Sanctuary, Western Ghats, India. Zoos' Print J. 2001;16(7):541–7.
83. Srinivasan U, Nuggihalli P. Additions to the avifauna of the Biligirirangan Hills, Karnataka. Indian Birds. 2005;1(5):103–4.
84. Ramesh BR. Flora of B. R. Hills. Pondicherry: French Institute of Pondicherry; 1989.
85. Coombes R. Hanumappa Sudarshan: the quiet reformer who has set up health services for 600 000 people. Br Med J. 2009;338(b):1100–1.
86. Morab SG. The Soliga of Biligiri Rangana Hills. Calcutta: Anthropological Survey of India; 1977.
87. Zvelebil K. The language of the Sholegas, Nilgiri area, South India. J Am Orient Soc. 1990;110(3):417–33.
88. Madegowda C. Traditional knowledge and conservation. Econ Polit Wkly. 2009;May:65–9.
89. Setty S, Bawa K, Ticktin T, Madegowda C. Evaluation of a participatory resource monitoring system for nontimber forest products: the case of amla (*Phyllanthus* spp.) fruit harvest by Soligas in South India. Ecol Soc. 2008;13(2):19–36.
90. Jadegowda M, Ramesh M. Empowerment of Soliga tribes. Leisa India. 2008;13–4.
91. Poulsen J. Impact of invasive species on biodiversity conservation and poor people's livelihoods. Assessment and Management of Alien Species that Threaten Ecosystems, Habitats and Species; 2001 Mar 12–16; Montreal, Canada: Secretariat of the Convention on Biological Diversity.
92. Setty S, Joseph G, Madegowda C, Raghunandan. Strengthening traditional agriculture and enhancement of livelihood security among Soligas at Biligiri Rangaswamy Temple Wildlife Sanctuary. In: Arunachalam V, editor. Participatory plant breeding and knowledge management for strengthening rural livelihoods. Chennai: M. S. Swaminathan Research Foundation; 2007. p. 215–6.

93. Zaraska N. Health behaviours of the Soliga tribe women. Kingston, ON: Queen's University; 1997.
94. Go I. A-11 state primary census abstract for individual scheduled tribe. New Delhi: Office of the Registrar General and Census Commissioner; 2001. [cited 2011 06–10-11], Available from http://censusindia.gov.in/Tables_Published/SCST/scst_main.html.
95. India Go. T 00-006: Number of literates and literacy rate by sex. New Delhi: Office of the Registrar General & Census Commissioner; 2001. [cited 2011 06–10-11], Available from: http://www.censusindia.gov.in/Tables_Published/A-Series/A-Series_links/t_00_006.aspx.
96. India Go. Sub-Districts area profile. New Delhi: Office of the Registrar General & Census Commissioner; 2001. Available from: http://www.censusindia.gov.in/Census_Data_2001/Census_Data_Online/Area_Profile/SubDistrict_Profile.aspx?cki=EVUSTZfgPoa.
97. India Go. Karnataka - data highlights: scheduled tribes. New Delhi: Office of the Registrar General & Census Commissioner; 2001. [cited 2011 06–10-11], Available from: http://censusindia.gov.in/Tables_Published/SCST/scst_main.html.
98. Hiremath A, Sundaram B. The fire-Lantana cycle hypothesis in Indian forests. Conserv Soc. 2005;3(1):26–42.
99. Kothari A, Rai N. Biligiri Tiger Reserve: include Soligas in conservation plan. Deccan Herald. 2011; 09/02/2011.
100. Atran S. Folk biology and the anthropology of science: cognitive universals and cultural particulars. Behav Brain Sci. 1988;21:547–609.
101. Evans N, Levinson S. The myth of language universals: language diversity and its importance for cognitive science. Behav Brain Sci. 2009;32(5):429.
102. Croft W. Typology and universals. 2nd ed. Cambridge: Cambridge University Press; 2003.
103. Greenberg J. Language universals: with special reference to feature hierarchies. The Hague: Mouton; 1966.
104. Burenhult N, Levinson S. Language and landscape: a cross-linguistic perspective. Lang Sci. 2008;30(2-3):135–50.
105. Enfield N, Majid A, Staden M. Cross-linguistic categorisation of the body: introduction. Lang Sci. 2006;28:137–47.
106. Berlin B, Kay P. Basic color terms: their universality and evolution. Berkeley, CA: University of California Press; 1969.
107. Dedrick D. Naming the rainbow: colour language, colour science, and culture. London: Kluwer Academic; 1998.
108. Lucy J. The linguistics of "color". In: Hardin CL, Maffi L, editors. Color categories in thought and language. Cambridge: Cambridge University Press; 1997. p. 320–46.
109. Wierzbicka A. Shape and colour in language and thought. In: Schalley AC, Khlentzos D, editors. Mental states, Language and cognitive structure, vol. 2. Philadelphia: John Benjamins; 2007. p. 37–60.
110. Davidoff J, Davies I, Roberson D. Colour categories in a stone-age tribe. Nature. 1999;398:203–4.
111. Kay P, Regier T. Language, thought and color: recent developments. Trends Cogn Sci. 2006;10(2):51–4.
112. Kay P, Regier T. Color naming universals: the case of Berinmo. Cognition. 2007;102(2):289–98.
113. Brown C. Mode of subsistence and folk biological taxonomy. Curr Anthropol. 1985;26(1):43–64.
114. Ereshefsky M. The poverty of the Linnean hierarchy. Cambridge: Cambridge University Press; 2001.
115. Mayden R. A hierarchy of species concepts: the denouement in the sage of the species problem. In: Claridge M, Dawah H, Wilson M, editors. Species: the units of biodiversity. London: Chapman and Hall; 1997. p. 381–424.
116. Agapow P, Bininda-Emonds O, Crandall K, Gittleman J, Mace G, Marshall J, et al. The impact of species concept on biodiversity studies. Q Rev Biol. 2004;79(2):161–79.

117. Cavalier-Smith T. Only six kingdoms of life. Proc R Soc Lond. 2004;271:1251–62.
118. Viruel J, Segarra-Moragues J, Perez-Collazos E, Villar L, Catalan P. Systematic revision of the *Epipetrum* group of *Dioscorea* (Dioscoreaceae) endemic to Chile. Syst Bot. 2010;35(1):40–63.
119. Brunner S. Fur seals and sea lions (Otariidae): identification of species and taxonomic review. Syst Biodivers. 2004;1(3):339–439.
120. Balakrishnan R. Species concepts, species boundaries and species identification: a view from the tropics. Syst Biol. 2005;54(4):689–93.
121. Walters S. Tansley Review 6 - The name of the rose: a review of ideas on the European bias in angiosperm classification. New Phytol. 1986;104:527–46.
122. Mitchell A. Systematic relationships of New Zealand endemic Brassicaceae inferred from nrDNA ITS sequence data. Syst Bot. 2000;25(1):98–105.
123. Walters S. The shaping of angiosperm taxonomy. New Phytol. 1961;60(1):74–84.
124. Clayton W. The genus concept in practice. Kew Bull. 1982;38:149–53.
125. Willis JC. The birth and spread of plants. Boissiera. 1949;8.
126. Cronk Q. The name of the pea: a quantitative history of legume classification. New Phytol. 1990;116(1):13–175.
127. Cronk Q. Measurement of biological and historical influences on plant classifications. Taxon. 1989;38(3):357–70.
128. Bartlett H. The concept of the genus: I. History of the generic concept in botany. Bull Torrey Bot Club. 1940;67(5):349–62.
129. DeQuieroz K. Ernst Mayr and the modern concept of species. Proc Natl Acad Sci USA. 2005;102 Suppl 1:6600–7.
130. DeQuieroz K, Gauthier J. Phylogenetic taxonomy. Annu Rev Ecol Syst. 1992;23:449–80.
131. Gaston K, May R. Taxonomy of taxonomists. Nature. 1992;356:281–2.
132. Younger B. The segregation of items into categories by 10-month-old infants. Child Dev. 1985;56:1574–83.
133. Vauclair J. Categorization and conceptual behavior in nonhuman primates. In: Bekoff M, Allen C, Burghardt GM, editors. The cognitive animal: empirical and theoretical perspectives on animal cognition. Cambridge: MIT Press; 2002.
134. Thompson R, Oden D. Categorical perception and conceptual judgments by nonhuman primates: the paleological monkey and the analogical ape. Cogn Sci. 2000;24(3):363–96.
135. Benard J, Stach S, Giurfa M. Categorization of visual stimuli in the honeybee *Apis mellifera*. Anim Cogn. 2006;9:257–70.
136. Avargues-Weber A, Portelli G, Benard J, Dyer A, Giurfa M. Configural processing enables discrimination and categorization of face-like stimuli in honeybees. J Exp Biol. 2010;213:593–601.
137. Nazzi T, Gopnik A. Linguistic and cognitive abilities in infancy: when does language become a tool for categorization? Cognition. 2001;80(3):B11–20.
138. Plunkett K, Hu J-F, Cohen L. Labels can override perceptual categories in early infancy. Cognition. 2008;106:665–81.
139. Cain AJ. The genus in evolutionary taxonomy. Syst Zool. 1956;5:97–109.
140. Stross B. Acquisition of botanical terminology by Tzeltal children. In: Edmonson M, editor. Meaning in Mayan languages. Hague: Mouton; 1973. p. 107–41.
141. Brown C. Linguistic ethnobiology: Amerindian oak nomenclature. In: Ford R, editor. Ethnobiology at the millennium: past promise and future prospects. Ann Arbor, MI: University of Michigan; 2001.
142. Stevens P. Review of *Ethnobiological Classification* - B. Berlin, 1992. Syst Biol. 1994;43(2):293–5.
143. Zarger R, Stepp J. Persistence of botanical knowledge among Tzeltal Maya children. Curr Anthropol. 2004;45(3):413–8.
144. Quinn P, Johnson M. Global-before-basic object categorization in connectionist networks and 2-month-old infants. Infancy. 2000;1(1):31–46.

145. Pauen S. The global-to-basic level shift in infants' categorical thinking: first evidence from a longitudinal study. Int J Behav Dev. 2002;26:492–9.
146. Stevens P. How to interpret botanical classifications - suggestions from history. Bioscience. 1997;47(4):243–50.
147. Hunn E. A Zapotec natural history. Tucson, AZ: The University of Arizona Press; 2008.
148. Si A. Aspects of honeybee natural history according to the Solega. Ethnobiol Lett. 2013;4:78–86.
149. Hunn E, French D. Alternatives to taxonomic hierarchy: the Sahaptin case. J Ethnobiol. 1984;4(1):73–92.
150. Hays T. Ndumba folk biology and general principles of ethnobotanical classification and nomenclature. Am Anthropol. 1983;85(3):592–611.
151. Bulmer R. Folk biology in the New Guinea Highlands. Social Sci Inf. 1974;13(4/5):9–28.
152. Bentley J, Rodriguez G. Honduran folk entomology. Curr Anthropol. 2001;42(2):285–300.
153. Hays T. Utilitarian/adaptationist explanations of folk biological classification: some cautionary notes. J Ethnobiol. 1982;2(1):89–94.
154. Harper D. A bioeconomic study of numeracy and economic calculation. J Bioecon. 2008;10:101–26.
155. Hurford J. Language and number: the emergence of a cognitive system. Oxford: Basil Blackwell; 1987.
156. Lean G. Counting systems of Papua New Guinea and Oceania. Lae: Papua New Guinea University of Technology; 1992.
157. Hammarstrom H. Number bases, frequencies and lengths cross-linguistically. Linguistic Perspectives on Numerical Expressions; 2004 June 10–11; Utrecht, Netherlands.
158. Ridley M. (Re)reading *The Origin*. Curr Biol. 2009;19(3):R2–3.
159. Hooke D. Trees in the Anglo-Saxon landscape: the Charter evidence. In: Biggam CP, editor. From earth to art: the many aspects of the plant-world in Anglo-Saxon England. Amsterdam: Rodopi; 2003. p. 17–40.
160. Gelling M. Place-names in the landscape. London: J. M. Dent; 1984.
161. Astour M. Ancient North Syrian toponyms derived from plant names. In: Rendsburg G, Adler R, Arfa M, Winter N, editors. The Bible world: essays in honor of Cyrus H. Gordon. New York: KTAV Publishing House, 1980. p. 1–8
162. Kislev M. Reference to the pistachio tree in Near East geographical names. Palest Explor Q. 1985;117(2):133–8.
163. Gyorffy E, editor. Old Hungarian river names in the multilingual Carpathian Basin. 23rd International Congress of Onomastic Sciences; 2008; Toronto: Youk University.
164. Baker B. 'I'm going to where-her-brisket-is': placenames in the Roper. In: Hercus L, Hodges F, Simpson J, editors. The land is a map: placenames of indigenous origin in Australia. Canberra: Pandanus Books and Pacific Linguistics; 2002. p. 103–29.
165. Ryan B. Some aboriginal place names on the mid-north coast of New South Wales. Oceania. 1964;34(4):278–307.
166. Si A, Agnihotri S. Solega placenames and their ecological significance. Anthropol Ling. 2014;56(3–4):389–414.
167. Agnihotri S, Si A. Solega ethno-ornithology. J Ethnobiol. 2012;32(2):185–211.
168. Grimmett R, Inskipp T. Birds of Southern India. London: A & C Black Publishers; 2005.
169. Diamond J, Bishop D. Ethno-ornithology of the Ketengban People, Indonesian New Guinea. In: Medin D, Atran S, editors. Folk biology. Cambridge, MA: MIT Press; 1999. p. 17–46.
170. Ng'weno F. Sound, sight, stories and science: avoiding pitfalls in ethno-ornithological research, with examples from Kenya. In: Tidemann S, Gosler A, editors. Ethno-ornithology: birds, indigenous peoples, culture and society. London: Earthscan; 2010. p. 103–14.
171. Grace G. The linguistic construction of reality. London: Routledge; 1987.
172. Wierzbicka A. Empirical universals of language as a basis for the study of other human universals and as a tool for exploring cross-cultural differences. Ethos. 2005;33(2):256–91.
173. Bulmer R. Mystical and mundane in Kalam classification of birds. In: Ellen R, Reason D, editors. Classifications in their social context. London: Academic Press; 1979. p. 57–79.

174. Bulmer R. Which came first, the chicken or the egg-head? In: Pouillon J, Maranda P, editors. Echanges et communications, melanges offerts a Claude Levi-Strauss, vol. 2. Paris/The Hague: Mouton; 1970. p. 1069–91.
175. Ghorpade K. Current species identities and scientific names of Indian birds: doubts, queries and counsel. Curr Sci. 2011;100(7):981–3.
176. Oiseaux.net. Family: Nectariniidae. Moulin de Bazoncourt: Écopains d'abord 2005 [29 August 2011]; Available from: http://www.oiseaux.net/birds/nectariniidae.html.
177. Peterson A, Navarro-Siguenza A. Consistency of taxonomic treatments: a response to Remsen (2005). Auk. 2006;123(3):885–7.
178. Ichikawa M. The birds as indicators of the invisible world: ethno-ornithology of the Mbuti hunter-gatherers. Afr Stud Monogr. 1998;25(Suppl):105–21.
179. Kizungu B, Ntabaza M, Mburunge M. Ethno-ornithology of the Tembo in Eastern DRC (former Zaire): Part one. Kalehe Zone. Afr Stud Monogr. 1998;19(2):103–13.
180. Hunn E. Sahaptin bird classification. In: Pawley A, editor. Man and a half. Auckland: The Polynesian Society; 1991. p. 137–47.
181. Goodale E, Kotagama S. Vocal mimicry by a passerine bird attracts other species involved in mixed-species flocks. Anim Behav. 2006;72:471–7.
182. Magrath R, Pitcher B, Gardner J. A mutual understanding? Interspecific responses by birds to each other's aerial alarm calls. Behav Ecol. 2007;18(5):945–51.
183. Finn F. Indian sporting birds. London: Francis Edwards; 1915.
184. Hume A. In: Oates E, editor. The nest and eggs of Indian birds. London: R. H. Porter; 1889.
185. Jones DB. Proteins of the castor bean - their preparation, properties and utilization. J Am Oil Chem Soc. 1947;24:247–51.
186. Logan C, Johnston T. Synthesis and separation in the history of "nature" and "nurture". Dev Psychobiol. 2007;49:758–69.
187. Wyman R. Experimental analysis of nature-nurture interactions. J Exp Zool A Comp Exp Biol. 2005;303A(6):415–21.
188. Nisbett R, Norenzayan A. Culture and cognition. In: Pashler H, Medin D, editors. Stevens' handbook of experimental psychology. 3rd ed. Hoboken, NJ: Wiley; 2002.
189. Dunn M, Greenhill SJ, Levinson SC, Gray RD. Evolved structure of language shows lineage-specific trends in word-order universals. Nature. 2011;473(7345):79–82.
190. Hallpike CR. The foundations of primitive thought. Oxford: Clarendon Press; 1979.
191. Mark D, Turk A, Burenhult N, Stea D. Landscape in language: an introduction. In: Mark D, Turk A, Burenhult N, Stea D, editors. Landscape in language: transdisciplinary perspectives. Amsterdam: John Benjamins; 2011.
192. Johnson L, Hunn E, editors. Landscape ethnoecology: concepts of biotic and physical space. New York: Berghahn; 2011.
193. Hunn E, Meilleur B. Toward a theory of landscape ethnoecological classification. In: Mann L, Hunn E, editors. Landscape ethnoecology: concepts of biotic and physical space. New York: Berghahn; 2011.
194. Gärling T, Book A, Lindberg E. Cognitive mapping of large-scale environments: the interrelationship of action plans, acquisition, and orientation. Environ Behav. 1984;16(3):3–34.
195. Leutgeb S, Leutgeb J, Barnes C, Moser E, McNaughton B, Moser M-B. Independent codes for spatial and episodic memory in hippocampal neuronal ensembles. Science. 2005;309:619–23.
196. McNaughton B, Battaglia F, Jensen O, Moser E, Moser M-B. Path integration and the neural basis of the 'cognitive map'. Nat Rev Neurosci. 2006;7:663–78.
197. Pocock D. Some characteristics of mental maps: an empirical study. Trans Inst Br Geogr New Ser. 1976;1(4):493–512.
198. Linde C, Labov W. Spatial networks as a site for the study of language and thought. Language. 1975;51(4):924–39.
199. Burrow T, Emeneau M. A Dravidian etymological dictionary. 2nd ed. Oxford: Clarendon Press; 1984.

200. Pawley A. Lexicalization. In: Tannen D, Alatis J, editors. Languages and linguistics: the interdependence of theory, data, and application. Washington, DC: Georgetown University Press; 1986.
201. Kull K. Ecosystems are made of semiosic bonds: consortia, umwelten, biophony and ecological codes. Biosemiotics. 2010;3(3):347–57.
202. Cobley P. The cultural implications of biosemiotics. Biosemiotics. 2010;3(2):225–44.
203. Kull K. Semiosphere and a dual ecology: paradoxes of communication. Sign Syst Stud. 2005;33(1):175–89.
204. Sebeok T. The sign science and the life science. Appl Semiot. 1999;3(6/7):85–96.
205. Eulau H. Perceptions of class and party in voting behavior: 1952. Am Polit Sci Rev. 1955;49(2):364–84.
206. Lo A, Reyes A. Language, identity and relationality in Asian Pacific America: an introduction. Pragmatics. 2004;14(2/3):115–25.
207. Clayton N, Dally J, Emery N. Social cognition by food-caching corvids. The western scrub-jay as a natural psychologist. Philos Trans R Soc. 2007;362(1480):507–22.
208. Cheney D, Seyfarth R, Smuts B. Social relationships and social cognition in nonhuman primates. Science. 1986;234(4782):1361–6.
209. Rist L, Shanker RU, Milner-Gulland E, Ghaznoul J. The use of traditional ecological knowledge in forest management: an example from India. Ecol Soc. 2010;15(1):3.
210. Tanaka J, Curran T. A neural basis for expert object recognition. Psychol Sci. 2001;12(1):43–7.
211. McShea D. Metazoan complexity and evolution: is there a trend? Evolution. 1996;50(2):477–92.
212. Gould S. Full house: the spread of excellence from Plato to Darwin. New York: Three Rivers Press; 1996.
213. McConvell P. The origin of subsections in Northern Australia. Oceania. 1985;56:1–33.
214. Fortunato L, Jordan F. Your place or mine? A phylogenetic comparative analysis of postmarital residence in Indo-European and Austronesian societies. Philos Trans R Soc B. 2010;365(1559):3913–22.
215. Gray R, Drummond A, Greenhill S. Language phylogenies reveal expansion pulses and pauses in Pacific settlement. Science. 2009;323(5913):479–83.
216. Dunnell R. The concept of progress in cultural evolution. In: Nitecki M, editor. Evolutionary progress. Chicago: The University of Chicago Press; 1988. p. 169–94.
217. Demmer U. Voices in the forest: the field of gathering among the Jenu Kurumba. In: Hockings P, editor. Blue mountains revisited: cultural studies on the Nilgiri Hills. Delhi: Oxford University Press; 1997. p. 165–91.
218. Kapp DB. Honigsammeln und Jagen bei den Alu Kurumbas. Anthropos St Augustin. 1983;78:715–38.
219. Campbell M. Busy bees: utopia, dystopia, and the very small. J Mediev Early Mod Stud. 2006;36(3):619–42.
220. Prete F. Can females rule the hive? The controversy over honey bee gender roles in British beekeeping texts of the sixteenth--eighteenth centuries. J Hist Biol. 1991;24(1):113–44.
221. Aristotle. Generation of animals, Book III. Cambridge, MA: Harvard University Press; 1953.
222. Aristotle. History of animals: In ten books. Tr. by Richard Cresswell. London: Bell; 1883.
223. Gedde J. The English apiary; or Compleat Bee Master. London: E. Curll, W. Mears, and T. Corbet; 1721.
224. Dyer F. Nocturnal orientation by the Asian honey bee *Apis dorsata*. Anim Behav. 1985;33:769–74.
225. Thanos C. Aristotle and Theophrastus on plant-animal interactions. In: Arianoutsou M, Groves RH, editors. Plant-animal interactions in Mediterranean-type ecosystems. Amsterdam: Kluwer Academic; 1994. p. 3–11.
226. Negbi M. Male and female in Theophrastus's botanical works. J Hist Biol. 1995;28(2):317–32.

227. Ellen R. Omniscience and ignorance: variation in Nuaulu knowledge, identification and classification of animals. Lang Soc. 1979;8(3):337–64.
228. Madock K. The emu anomaly. In: Hiatt L, editor. Australian aboriginal mythology. Canberra: Australian Institute of Aboriginal Studies; 1975.
229. Wierzbicka A. Understanding cultures through their key words. New York: Oxford University Press; 2002.
230. Baird J, Wagner M, Noma E. Impossible cognitive spaces. Geogr Anal. 1982;14(3):204–16.
231. Tversky B. Distortions in cognitive maps. Geoforum. 1992;23(2):131–8.
232. Ellen R. In: Gellner E, Goody J, Gudeman S, Herzfeld M, Parry J, editors. The cultural relations of classification: an analysis of Nuaulu animal categories from Central Seram. Cambridge: Cambridge University Press; 1993.
233. Ellen R, Reason D, editors. Classifications in their social context. London: Academic Press; 1979.
234. Evans N. Kayardild dictionary and thesaurus. Melbourne: University of Melbourne; 1992.
235. Hunn E. What is traditional ecological knowledge? In: Williams N, Baines G, editors. Traditional ecological knowledge: wisdom for sustainable development. Canberra: Australian National University; 1993.
236. Lewis H. Traditional ecological knowledge: some definitions. In: Williams N, Baines G, editors. Traditional ecological knowledge: wisdom for sustainable development. Canberra: Australian National University; 1993.

Index

A. Si, *The Traditional Ecological Knowledge of the Solega*, Ethnobiology,
DOI 10.1007/978-3-319-24681-9

MIX
Papier aus verantwortungsvollen Quellen
Paper from responsible sources
FSC® C105338

If you have any concerns about our products,
you can contact us on
ProductSafety@springernature.com

In case Publisher is established outside the EU,
the EU authorized representative is:
Springer Nature Customer Service Center GmbH
Europaplatz 3, 69115 Heidelberg, Germany

Printed by Libri Plureos GmbH
in Hamburg, Germany